Markt- und Unternehmensentwicklung / Markets and Organisations

Edited by
A. Picot, München, Deutschland
R. Reichwald, Leipzig, Deutschland
E. Franck, Zürich, Schweiz
K. M. Möslein, Erlangen-Nürnberg, Deutschland

Change of institutions, technology and competition drives the interplay of markets and organisations. The scientific series 'Markets and Organisations' addresses a magnitude of related questions, presents theoretic and empirical findings and discusses related concepts and models.

Edited by

Professor Dr. Dres. h. c. Arnold Picot
Ludwig-Maximilians-Universität
München, Deutschland

Professor Dr. Professor h. c. Dr. h. c.
Ralf Reichwald
HHL – Leipzig Graduate School
of Management, Leipzig, Deutschland

Professor Dr. Egon Franck
Universität Zürich, Schweiz

Professorin Dr. Kathrin M. Möslein
Universität Erlangen-Nürnberg,
Deutschland,
HHL – Leipzig Graduate School
of Management, Leipzig, Deutschland

Stefan H. Hallerstede

Managing the Lifecycle of Open Innovation Platforms

Springer Gabler

Stefan H. Hallerstede
Lehrstuhl Wirtschaftsinformatik I
Universität Erlangen-Nürnberg
Nürnberg, Germany

Dissertation Universität Erlangen-Nürnberg, 2013

ISBN 978-3-658-02507-6   ISBN 978-3-658-02508-3 (eBook)
DOI 10.1007/978-3-658-02508-3

The Deutsche Nationalbibliothek lists this publication in the Deutsche Nationalbibliografie;
detailed bibliographic data are available in the Internet at http://dnb.d-nb.de.

Library of Congress Control Number: 2013938287

Springer Gabler
© Springer Fachmedien Wiesbaden 2013

Printed on acid-free paper

Springer Gabler is a brand of Springer DE.
Springer DE is part of Springer Science+Business Media.
www.springer-gabler.de

Foreword

Open innovation denotes innovation processes that are not solely bound to R&D departments. Open innovation processes cross the boundaries of organizational units and organizations and involve new types of players in innovation activities. Open innovation, thus, expands the solution space for new products and processes, services and solutions. Independent and distributed actors who are no longer bound to an organization can serve as idea providers, concept developers or implementers of innovations. Open innovation platforms facilitate and support those processes in a virtual environment. They provide users with IT-based tools for open innovation such as innovation communities, innovation contests, innovation toolkits or innovation marketplaces. While the phenomenon of open innovation, its central actors, the specific processes, projects, and products has been investigated for more than a decade, the design and management of the underlying platforms have received much less attention until now.

Stefan Hallerstede identified this highly relevant gap in research and business practice and dedicated his efforts to shed light on core questions in the field. The idea to connect current research in information systems with research in open innovation represents a convincing approach to advance, in particular, the research in open innovation. It also adds a new level of professionalism to the field.

The work of Stefan Hallerstede focuses on managing the lifecycle of open innovation platforms and addresses the following two major research questions:

- Who should manage the lifecycle of open innovation platforms?

- What is the process of open innovation platform lifecycle management?

Based on exploratory research, the contribution offers valuable insights into three relevant aspects concerning the lifecycle management of open innovation platforms: the platforms themselves, their lifecycle managers, and a model to structure their lifecycle. All three aspects are discussed in detail and implications are derived. The topics Stefan Hallerstede investigates and the results he presents are highly relevant for the successful realization of IT-based open innovation projects. From a research perspective, the focus of the study is well-chosen, and the contribution substantial. Its style captivates the reader from the very first to the very last page. The book is accessibly written and due to its well thought-out didactic structure, one will easily learn of interesting details and gain important impulses in the course of reading. In conclusion, the work appeals by its practical reach, academic scope, and fresh perspective. It has been accepted as doctoral dissertation in 2013 by the School of Business and Economics at the University of Erlangen-Nuremberg. The book is a must-read for all those who intent to use – or are already using – IT-based tools for open innovation: researchers as well as practitioners. I wish the book the broad dissemination it deserves.

Prof. Dr. Kathrin M. Möslein

Preface

I have always been fascinated by online platforms and by bringing them to life. Finally, I have found a way to turn my passion into profession. Similar to what occurs in the field of open innovation, this thesis would not have been possible without the support of many sources. Therefore, I am indebted to a great number of persons. I would like to name the most influencing ones, but at the same time dearly thank everyone who has helped me in one way or another.

My foremost thanks goes to my supervisor Prof. Dr. Kathrin M. Möslein, head of the chair of information systems I of the University of Erlangen-Nuremberg, for letting me be a part of her team and providing me the great opportunity to write a thesis under her guidance. Dr. Angelika C. Bullinger oversaw the process from my very first day as a doctoral student. Her support and ideas helped me to identify the research field, to develop the research design, to execute it and to finally publish results, which we did collaboratively many times. My dear colleagues Christiane Rau and Uta Renken challenged all my decisions and words in order to further sharpen each and every single argument of this work. Not shaping arguments but language, Barbara Hentschel was the source of raising this thesis to a new level of style.

But what is style and excellence without content? This research would not have been possible without the case companies, namely Atizo, HYVE and innosabi. I gratefully acknowledge their openness, cooperation and unreserved willingness to let me analyze their business.

Besides these main contributors, there are many others who supported me through constructive input. To name only two, I would like to express my gratitude to Prof. Dr. Michael Amberg for the evaluation of this thesis as its second supervisor and to Prof. Dr. Dr. h.c. mult. Peter Mertens for providing me with small comments, hints, and motivations whenever our paths crossed in the hallway.

Research, and hence this thesis, is more and more linked to the research community spread all over the world. I wish to thank the German Federal Ministry of Education and Research (project: BALANCE, FKZ 01FH09153) in particular for funding the dissemination and discussion trips to various national and international conferences. They enabled me to meet many outstanding researchers who helped to improve this work by pointing me towards yet unrecognized directions.

Finally, special thanks goes to all team members of the chair for information systems I for official and unofficial, fruitful and unfruitful as well as day- and night-time discussions on research.

Research is often considered a painful and tedious process. I think this is true if you have to face it on your own. I feel privileged to have had so many companions along the way. Especially my family and friends were extremely patient and supportive with regard to my work-life-research-balance, which I gratefully acknowledge. They all together made my research journey rather enjoyable than painful.

I would do it again.

Stefan H. Hallerstede

Overview of contents

Table of contents

List of figures

List of tables

List of abbreviations

ALM	Application lifecycle management
CAD	Computer-aided design
CEO	Chief executive officer
CFO	Chief financial officer
CMO	Chief marketing officer
CTO	Chief technical officer
GfdS	Gemeinsam fuer die Seltenen
HIC	HYVE Innovation Community
HID	HYVE Innovation Design
HIR	HYVE Innovation Research
IS	Information systems
IT	Information technology
ITIL	Information technology infrastructure library
MVC	Model view controller
NIH	Not-invented-here
OCLM	Online community lifecycle model
OGC	Office of Government Commerce
OI	Open innovation
OII	Open innovation intermediary
OIP	Open innovation platform
OIP-LM	OIP lifecycle model
ORM	Object relational mapper
R&D	Research and development
RUP	Rational unified process
SaaS	Software-as-a-service
STS	Socio-technical system
UML	Unified modeling language
XP	Extreme programming

Abstract

Innovation is a critical success factor for organizations to survive. Nowadays, possibilities of information and communication technologies facilitate the use of IT-based tools for the integration of external innovators into the innovation process of organizations. These tools are subsumed under the term *open innovation platform* (OIP). Although many insights concerning OIPs are available, a dedicated approach to manage an OIP throughout its lifecycle is so far missing. However, the skills required to design and manage OIPs differ from those necessary for default websites. Thus, a dedicated lifecycle approach is vital for an OIP's success. This contribution develops a *lifecycle management approach for OIPs* based on *three in-depth cases* of professional OIP lifecycle management. Valuable insights are provided for both organizers and lifecycle managers of OIPs.

According to the OIP lifecycle model, which is developed in this thesis, an OIP lifecycle consists of *OIP design* and *OIP management*. Whereas OIP design refers to the phases *requirements*, *design* and *build*, OIP management refers to the phases *deploy*, *operate* and *optimize*. The reader learns which phases and activities are crucial in an OIP lifecycle, which ones can be neglected, and which factors influence an appropriate OIP lifecycle management. The major challenges in an OIP lifecycle are discussed, i.e. selecting the right problems, formulating them, overcoming employees' reluctance as well as facilitating software-mediated knowledge transfer. It is shown how the challenges can be addressed and leveled. In addition, the present study introduces a process to design the socio-technical system of an OIP that acknowledges the complexity and importance of an OIP's social subsystem, like its motivational system.

It is shown that *open innovation intermediaries* (OIIs), as professional OIP lifecycle managers, fulfill certain *functions* for organizers of OIPs. However, OIIs currently fall back in services during OIP management, which offers potential to differentiate themselves from their competitors. This thesis presents benefits and drawbacks of *four types of OIP projects* derived from the dimensions of OIP design and OIP management. OIIs are characterized by one type of OIP project. According to these findings, organizers can select an appropriate OII based on the intended type of OIP project.

Aside from the implications for practitioners, *researchers* benefit from the structuring of the field as well as from the merge of open innovation and information systems literature. Future research needs are identified.

Part I

Introduction

1 Motivation and relevance

Innovate or die! (Thomas J. Peters)

Thomas J. Peters[1] expresses the challenge corporations are confronted with if they want to ensure their survival: they have to innovate or die[2]. Innovation[3] as a critical success factor for organizations was already recognized by Schumpeter[4] in the 1930s and constantly gained importance since then. It is particularly important in today's dynamic markets[5]. What has evolved significantly in the recent past is the *way* innovation is generated[6]. Higher competitive pressure and increasingly shorter product lifecycles forced companies to develop new strategies to avoid death[7]. Knowledge from diverse domains had to be combined rapidly in order to create competitive innovations. Traditional research and development departments were not able to keep up with these developments on their own[8]. These changes caused companies to realize the potential of integrating external innovators, such as customers, consumers, suppliers, and even competitors, into their innovation processes[9]. For this new paradigm, Henry Chesbrough coined the term open innovation[10].

Nowadays, possibilities of information and communication technologies, such as the internet, facilitate the use of IT-based tools for open innovation[11] like innovation contests[12], innovation communities[13], innovation market places[14] and

---

[1] Thomas J. Peters is a business author, consultant and firm owner in the field of management. See Crainer (2001).
[2] See also Peters and Waterman (1982).
[3] Innovation is defined as the successful commercial exploitation of an invention. See OECD (1991).
[4] Schumpeter (1934).
[5] Christensen (2006).
[6] Von Hippel (2005).
[7] Vahs and Burmester (2005).
[8] Christensen (2006).
[9] Williams, Gownder, Wiramihardja and Corbett (2010). For a definition of the innovation process see Part II.1.2.
[10] Chesbrough (2003).
[11] Gassmann, Enkel and Chesbrough (2010); Hrastinski, Kviselius, Ozan and Edenius (2010); Moeslein and Neyer (2009).
[12] Bullinger and Moeslein (2011).
[13] Fueller, Bartl, Ernst and Muehlbacher (2006).
[14] Bakos (1997); Lichtenthaler and Ernst (2008).

innovation toolkits[15], for the integration of innovators[16]. In this thesis, these IT-based tools for open innovation are subsumed under the term *open innovation platform* (OIP)[17].

Integrating innovators into an organization's innovation process using open innovation platforms has become a popular means in business practice during the last two decades[18]. Johann Fueller is researcher in the field of open innovation and CEO[19] of the HYVE AG, which is one of the market leading consultancies for open innovation in Germany [20]. He puts it this way:

> *Back in those days when we started our journey, we met people who wanted to do something cool and new. They wanted to get beyond traditional market research to generate innovations. And we tried it out together with them. [...] Nowadays, open innovation resounds throughout the country and is accessible for everybody. It has become a standard in innovation management and in the meantime we know better how to do it. (Johann Fueller[21])*

In accordance with the rising popularity of open innovation, researchers in the field have started to investigate different aspects concerning open innovation platforms. They focus for instance on how to manage innovation communities [22], how to motivate its participants[23] or how to evaluate and select contributions[24]. Furthermore, they investigate the economic impact of open innovation platforms[25] and the design elements of innovation contests[26]. In sum, there is a great variety of research from different perspectives on open innovation platforms.

However, we know very little about linking all these findings. In other words, there is no approach to manage an open innovation platform throughout its lifecycle

---

[15] Von Hippel and Katz (2002).
[16] Terwiesch and Xu (2008); Williams et al. (2010).
[17] Open innovation platforms are defined and further characterized in Part II.1. Innovation technologies (see Part II.1.3.5), as another tool for IT-based open innovation are not subsumed under the term OIP.
[18] Bullinger and Moeslein (2011).
[19] Chief executive officer.
[20] Williams (2011).
[21] This quote is taken from a personal interview with Johann Fueller in 2011. The interview was conducted in course of data elicitation for this thesis. For more details, refer to Part III.1.3.
[22] Adamczyk, Bullinger and Moeslein (2010); Klein and Lechner (2009).
[23] Harhoff (2003); Walcher (2007).
[24] Moeslein, Haller and Bullinger (2010); Piller and Walcher (2006).
[25] Bishop (2009).
[26] Bullinger and Moeslein (2011).

from the initial idea to its discontinuation[27]. Iriberri and Leroy[28] address this topic in their work on innovation communities. They derive a lifecycle model for innovation communities and define success factors for single phases within the lifecycle. However, the authors acknowledge that further research is necessary to provide guidelines on how to *implement* these success factors by defining a development process for OIPs. In another line of argumentation, Gassmann et al.[29] request to add a more professional perspective to current research on open innovation platforms. In their paper on the future of open innovation, they recommend for instance introducing management processes for OIPs. Professionalizing the development of open innovation platforms is viable for both, practice and research, as the following two examples illustrate:

Currently, *practitioners* in the field have to rely on trial-and-error approaches and their experience to manage the lifecycle of open innovation platforms[30]. Using OIPs to integrate innovators without clear guidelines does not always lead to success, as the case of Henkel[31] shows: In 2011, Henkel announced an innovation contest to design a new bottle for its dishwashing detergent Pril. The innovation contest ended in a social media disaster for Henkel, when an intentionally poorly designed bottle with the idea of chicken flavored detergent won the peer evaluation. Henkel had to alter the contest rules in order to eliminate the design[32]. With an informed approach, this flaw could have been identified prior to the innovation contest's go-live.

Parallel to the lack of guidance for practitioners, *researchers* lack a structuring of the field from a lifecycle perspective. There is an ongoing discussion on the comparative importance of implications from prior research: Is it more important to take care of a user-centered design process during development, as for instance Leimeister and Krcmar[33] claim, or is it more important to invest time in community management, as for instance Zhang and Hiltz[34] claim? Due to this knowledge gap, it seems difficult to decide upon the emphasis for future research. Moreover,

---

[27] See Hrastinski et al. (2010). For a definition of an OIP lifecycle see Part II.3.
[28] Iriberri and Leroy (2009).
[29] Gassmann et al. (2010).
[30] Gassmann et al. (2010).
[31] www.henkel.de; retrieved December 3, 2012.
[32] Spiegel Online (2011).
[33] Leimeister and Krcmar (2004). This call is also supported by for instance Andrews, Preece and Turoff (2002); Cothrel and Williams (1999); Preece and Maloney-Krichmar (2006).
[34] Zhang and Hiltz (2003). This call is also supported by for instance Adamczyk (2012); Beenen et al. (2004); Hall and Graham (2004); Kim (2000).

structuring the field helps researchers to identify further research needs and practitioners to draw concertedly on research for particular phases in an OIP's lifecycle.

As stated above, open innovation research does not offer suitable models that might serve to structure the lifecycle of OIPs. However, the information systems (IS) discipline created multiple models for software lifecycle management since the 1980s and continuously improved them[35]. The idea of this thesis is to transfer findings from the information systems discipline to the field of open innovation with the ultimate target of professionalizing OIP lifecycle management. While open innovation research is strong in the field of IT-based tools for open innovation, IS research is strong in formal approaches for software lifecycle management. By bridging the two disciplines, I will contribute to both, information systems and open innovation research: *IS research* will benefit from additional details and cases on software lifecycle management. This will help to explain a software lifecycle model in more detail and characterize additional fields of its application. In the field of *open innovation research*, my aim is to increase the knowledge on managing the lifecycle of open innovation platforms and contribute to structuring and professionalizing it.

The remainder of this part is structured as follows. *Chapter 2* translates the objectives of this thesis into research questions. *Chapter 3* introduces the research design that is applied to address the research questions. Finally, *chapter 4* introduces the detailed structure of this thesis.

---

[35] Scacchi (2001).

2 Research questions

The aim of this thesis is to explore lifecycle management for open innovation platforms in order to reach a better understanding and to provide guidelines to manage the lifecycle of OIPs. The following consideration helps to break down the objective into research questions: On the one hand, there are many stories of success when organizations use open innovation. Companies call for solutions to problems they were not able to solve themselves and receive multiple breakthrough innovations on short notice[36]. On the other hand, as described above, the Henkel case shows that an OIP's lifecycle has to be managed appropriately in order to prevent failure. However, it is still unclear what "managed appropriately" means and who is capable of carrying out this task. Accordingly, two areas of interest arise: *Who* should manage the lifecycle of OIPs and *how* should it be done? The resulting research questions are set out in detail below.

The rising popularity and distribution of open innovation fostered the emergence of innovation intermediaries, who specialized on integrating innovators in organizers'[37], i.e. their clients', innovation processes using OIPs[38]. These specialized innovation intermediaries are called open innovation intermediaries (OIIs)[39]. The question arises if they are the right players to manage an OIP's lifecycle, and which *skills* would they have to display? Investigating this line of argumentation, prior research showed that open innovation intermediaries can offer different services (*functions*) to organizers of OIPs[40]. It is thus questionable whether the services of open innovation intermediaries in managing OIP lifecycles differ as well, and whether this influences the selection of an appropriate open innovation intermediary for a particular OIP project[41]. These problems are addressed by research question I):

---

[36] See for instance Diener and Piller (2010); Terwiesch and Xu (2008); www.atizo.com/success-page; retrieved October 22, 2012.

[37] An organizer of an OIP is for instance a company that wants to run an OIP projects. The organizer initiates an OIP project and might ask an OII to take care of OIP lifecycle management. For details see Part II.1.2.

[38] Diener and Piller (2010); Hossain (2012); Verona, Prandelli and Sawhney (2006).

[39] Jaervi, Schallmo and Kutvonen (2011) use this term in a broader sense to describe innovation intermediaries that apply open innovation methods in general. However, in this thesis, OIIs refer only to innovation intermediaries that use OIPs.

[40] See Lopez-Vega and Vanhaverbeke (2009).

[41] An OIP project is defined as all activities that are required to design and manage OIPs.

I)   Who should manage the lifecycle of open innovation platforms?

The second research question shifts the perspective from an organizer, who wants to run an OIP and thus needs to select an OIP lifecycle manager, towards the latter one. OIP lifecycle managers need guidance on how to manage an OIP's lifecycle. They cannot use existing software lifecycle models as they are, because such models have to be adapted to the type of software they refer to[42]. Thus, an OIP lifecycle manager lacks guidance on OIP lifecycle management[43]. In order to address this gap in research, research question II) suggests developing a model for OIP lifecycle management:

II)   What is the process of open innovation platform lifecycle management?

The following chapter explains how the two research questions are to be tackled in this thesis.

---

[42] Ahmad, Li and Azam (2005); Escalona and Koch (2003); Fraternali (1999); Orlikowski, Yates, Okamura and Fujimoto (1995); Pressman (1998). A more detailed reasoning can be found in Part II.3.6.

[43] See also Gassmann et al. (2010).

3 Research design

The aim of this thesis is to *explore* the yet unexplored field of *lifecycle management for OIPs*. Accordingly, an explorative research approach is chosen[44]. As the major questions concern the "who" and "how" of OIP lifecycle management[45], case study research is selected as research method[46]. The paradigm and structure of this thesis follow a linear-analytic multiple case study design according to Yin[47].

The following provides a brief overview of the research method and the research field. It introduces the socio-technical systems theory as a theoretical basis of this thesis, explains the contributions of each part, outlines the links between the parts and closes with a remark on the philosophy of this thesis.

To explore the "who" and "how" of OIP lifecycle management and factors involved, *in-depth case analyses of OIP projects* are conducted. *Open innovation intermediaries* potentially hold professional experience in processing OIP projects, as this is their daily business[48]. Condensing insights from multiple OIIs yields the opportunity of exploring patterns of professional OIP lifecycle management, being part of an OIP project. Thus, OIIs provide a suitable research field and are, therefore, selected as the research objects in the case study of this thesis. The primary data source for the case study are semi-structured explorative interviews[49].

Though being explorative, this thesis builds on prior research in the field of information systems and open innovation. The insights from prior research help to structure the exploration. Firstly, a software lifecycle model stemming from IS research is adapted to the field of open innovation. The adapted model constitutes the *OIP lifecycle model*, which is used to structure the case analyses and other considerations concerning the lifecycle of OIPs.

---

[44] Robson (2002); Schnell, Hill and Esser (2008).
[45] See chapter 2.
[46] See Benbasat, Goldstein and Mead (1987); Borchardt and Goethlich (2007); Creswell (2007); Yin (2009). The selection of case study research as a research method is further justified in Part III.1.1.
[47] Yin (2009). Linear-analytic case studies follow a common structure, which is also reflected in this thesis: problem, related work, methods, analysis, conclusions.
[48] See Diener and Piller (2010).
[49] Robson (2002).

Secondly, this thesis draws on the *socio-technical systems (STS) theory*[50], as OIP lifecycle management concerns social factors, e.g. the individuals involved in an OIP project, as well as technical factors, e.g. the processes involved in an OIP project. Bostrom and Heinen[51] used the STS theory to describe the management of information systems depending on multiple components. The implication for this thesis is that OIP lifecycle management has to be investigated in the light of social as well as technical components, their interdependence, input and output factors as well as its environment[52].

Accordingly, an OIP lifecycle manager that runs an OIP project has to consider the stakeholders of an OIP (*people*; e.g. users, project members, organizers), their relations to each other (*structure*; e.g. hierarchies, incentive systems), the tools supporting the OIP lifecycle management (*technology*; e.g. software lifecycle models, other processes, tools for communication) as well as the objective that is pursued (*task*; e.g. the aim to run a successful OIP project). While jointly optimizing the several components of the socio-technical system[53], an OIP lifecycle manager has to keep in mind the *input* factors to the STS (e.g. the knowledge of project members) that create a certain *output* (e.g. the successful OIP project) in the context of the STS's *environment* (e.g. budget constraints, legislation).

The initially stated question, whether it is more important to take care of a user-centered design process during development or whether it is more important to invest time in community management can be answered jointly using the socio-technical systems perspective. Thus, the socio-technical systems theory builds the overall theoretical basis for this thesis and is consequently inherent in all considerations. In other words, *OIP lifecycle management is researched in the light of the socio-technical systems theory*. The following introduces the research design to address the research questions.

Part I sets the stage by defining the *research questions and structure* for this thesis and outlining the importance of the present thesis. Furthermore, three relevant facets in OIP lifecycle management were identified[54]: (1) the research field, i.e. the OIPs, (2) the OIP lifecycle managers, i.e. the open innovation intermediaries, and (3) an

---

[50]   Trist and Bamforth (1951).
[51]   Bostrom and Heinen (1977).
[52]   The socio-technical systems theory is introduced in more detail in Part II.1.4.
[53]   Pasmore et al. (1982).
[54]   See chapter 1 and 2.

approach to structure the lifecycle of OIPs, i.e. software lifecycle models. Figure 1 depicts this relationship. Innovators who participate in an OIP as well as the organizers of OIPs (i.e. the OIIs' clients) are not addressed explicitly by the research design, as they are not part of the research questions. However, acknowledging the socio-technical systems theory, both parties are relevant since they have to be considered as users of an OIP (i.e. people).

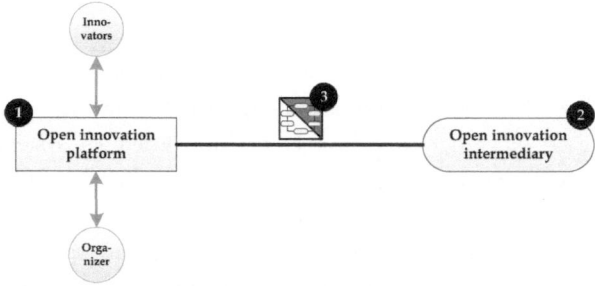

Figure 1: Three facets of OIP lifecycle management[55]

Part II sets the *foundations* for these three facets of OIP lifecycle management, as shown in Figure 2. Firstly, *open innovation platforms* are defined and characterized. As outlined in chapter 1, OIP is a term that subsumes IT-based tools for open innovation. The objective of the case study is to cover all[56] IT-based tools for open innovation by at least one case. Consequently, the case selection for the case study is informed by the IT-based tools for open innovation implemented on an OIP. This ensures comprehensiveness of the assessment and facilitates the generalization of findings. Furthermore, aiming at a better understanding of OIPs and their relevant components, OIPs are characterized based on the socio-technical systems theory. Applying STS theory ensures that no major component of an OIP is omitted in the analysis.

Secondly, *open innovation intermediaries*, who emerged as lifecycle managers for OIPs[57], are introduced. In order to set foci in the case study, functions of open innovation intermediaries and challenges of working with them in a virtual innovation environment are outlined from a literature perspective.

---

55 If no reference is provided for a figure or table, it is an own depiction. This notion is omitted in the following.
56 One of the five IT-based tools for open innovation is excluded from the analysis, as it deals mainly with hardware and not with software. For details see Part II.1.3.6.
57 Diener and Piller (2010).

Finally, a *software lifecycle model* is sought to structure the case analyses in order to ensure a comprehensive assessment of an OIP's lifecycle. This thesis draws on IS models, as information systems research developed multiple models for professional software lifecycle management. Requirements that a model should meet in order to be suitable for structuring the case study analysis are derived. Based on these requirements, multiple software lifecycle models are compared in order to select one. The selected model will be adapted to the field of OIPs, resulting in the OIP lifecycle model.

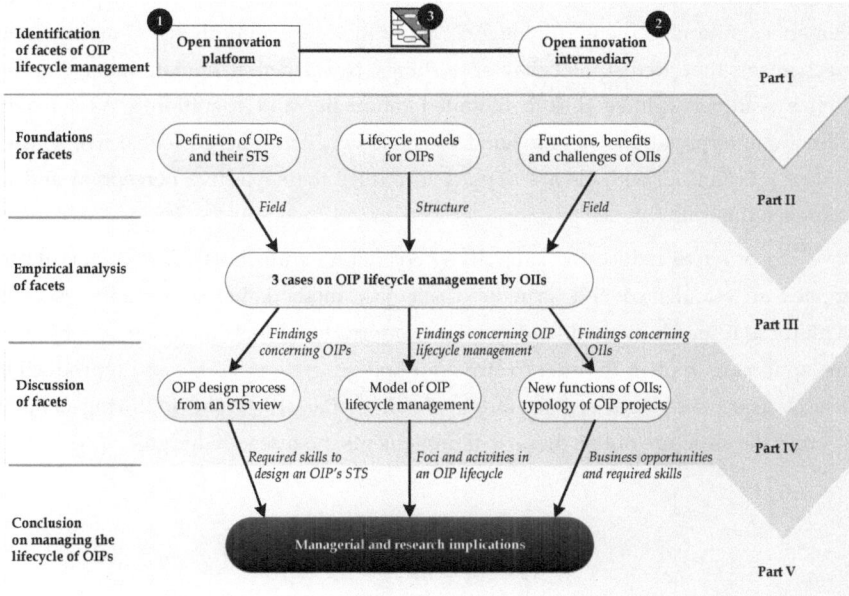

Figure 2: Research design

Part III describes findings from three in-depth *cases on OIP lifecycle management* by open innovation intermediaries. The case descriptions are structured along the findings of Part II. Firstly, the OII is introduced, secondly, the OIP is sketched, and thirdly, the OIP lifecycle management is examined. Each case closes with a conclusion concerning the single case.

Part IV discusses the findings from the single cases in a *cross-case analysis* in order to derive integrated findings. The aim is to answer the research questions of who should manage an OIP's lifecycle and how it should be done. Relating and integrating the findings of all cases and comparing it with prior research expands the knowledge on OIP lifecycle management and on open innovation intermediaries. Thereby, the structure of the cross-case analysis is informed by the foundations of Part II. Major aspects of each facet in OIP lifecycle management, i.e. the OII, the OIP and the lifecycle model, are discussed.

Part V closes the thesis and draws *managerial and research implications* based on the findings[58]. Contributions further the knowledge in the fields of open innovation and information systems.

The present thesis and its research design follow the concept of *critical realism*[59]. According to critical realism, there is an objective, real world, which is independent of human cognition (objective ontology). Thus, there are underlying structures and mechanisms that predict the behavior of things. Nevertheless, human understanding of this reality is limited due to bounded rationality and cognition[60]. As a result, different interpretations of the same phenomenon can emerge (subjective episte- mology). Scientific methods are applied to reduce the subjective perception and to approach the truth.

In order to reduce the subjective perception as much as possible, this thesis applies an established and structured scientific method, i.e. case study research. Within the research method, dedicated procedures to address the issue of subjective perception are applied in order to strive for and generate traceable and reproducible results. The procedures are set out in detail in Part III.1. The following chapter outlines the structure of this thesis that implements the research design.

---

[58] Trinder and Reynolds (2000).
[59] Bhaskar (2008); Collier (1994); Ven (2007).
[60] Simon (1957).

4 Structure of thesis

This work is divided into five parts. The structure reflects a linear-analytic case study design according to Yin[61]. The *parts* are broken down into *chapters*, which are further split into *sections* and *subsections*. The five parts build on each other and should, therefore, be read subsequently. For a quick reading a short summary of each major part will be provided at its end. An overview of the five parts, as summarized in Figure 3, follows.

Part I - Introduction sets the stage by highlighting the relevance of the research topic (*chapter 1*), the research gap and the thereof derived research questions (*chapter 2*). Secondly, it introduces the overall research design (*chapter 3*) and structure of the thesis (*chapter 4*).

Part II - Foundations sheds light on the three facets that are relevant to managing the lifecycle of open innovation platforms, as identified in Part I: open innovation platforms, open innovation intermediaries and lifecycle models. As this thesis aims at bridging information systems and open innovation literature[62], each chapter in Part II begins with a short summary of the research streams that are relevant to that chapter in order to outline the origin of the foundations. Firstly, the term open innovation platform is introduced as a cover term for IT-based tools for open innovation. Having characterized OIPs in a first step, their socio-technical system is analyzed in step two (*chapter 1*). Secondly, open innovation intermediaries, who serve as a research object, are introduced as specialized innovation intermediaries that manage the lifecycle of OIPs. Functions of OIIs as well as benefits and challenges of working with them are outlined (*chapter 2*). The last chapter in Part II addresses software lifecycles models. The need for a lifecycle approach for OIPs is outlined and a model to structure the case study is selected and adapted to the field, resulting in the OIP lifecycle model (*chapter 3*). The part closes with a short summary of the major findings (*chapter 4*).

Part III - Empirical study constitutes the research data of this work. Firstly, the research method, namely the case study design, is described (*chapter 1*). Secondly, findings from three cases of OIP lifecycle management by open innovation

---

[61] Yin (2009).
[62] See chapter 1.

intermediaries are portrayed. The cases are structured according to the three facets of an OIP's lifecycle, as identified in Part II: The open innovation intermediary, its OIP and its OIP lifecycle management are described in detail. Thereby, the individual ways to manage an OIP's lifecycle are highlighted and briefly discussed in order to reveal the OII's core competencies as well as good-practice examples of OIP lifecycle management (*chapters 2 to 4*). The part closes with a summary of findings from the three cases (*chapter 5*).

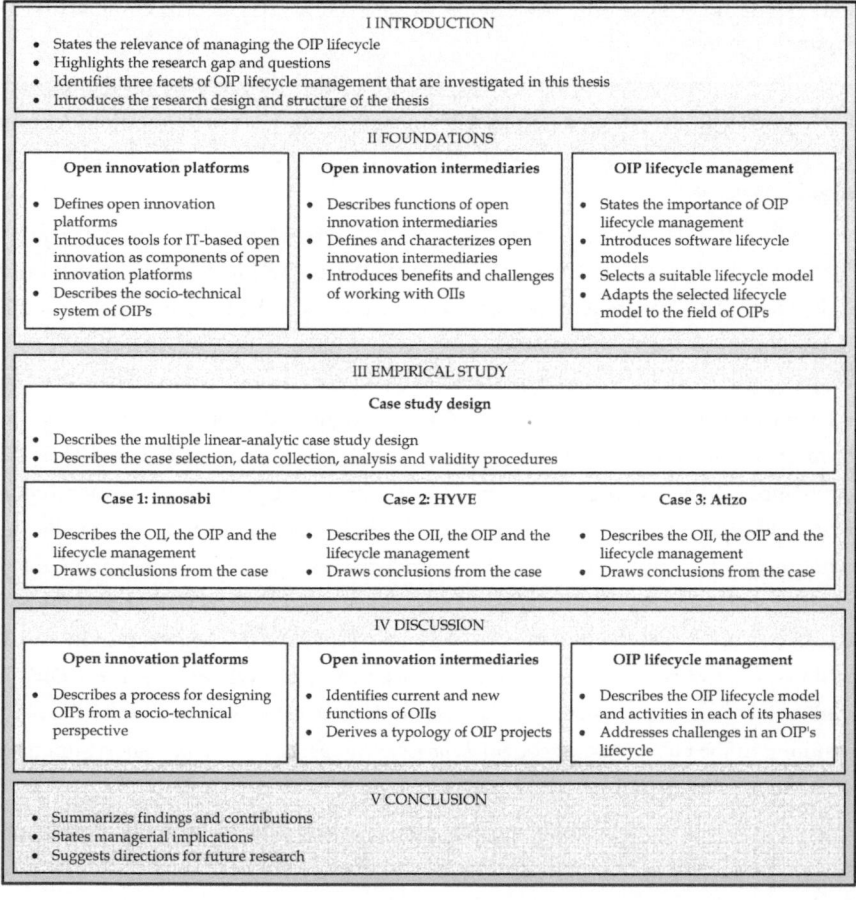

Figure 3: Structure of thesis

Part IV - Discussion merges findings from the single cases introduced in Part III in a cross-case analysis and contrasts the integrated findings with findings from the literature. The cross-case analysis is structured along the insights of the case descriptions in Part III. In the beginning, open innovation intermediaries are focused on: The first chapters reflect on the functions of open innovation intermediaries (*chapter 1*) and derive a typology of the OIP projects OIIs run (*chapter 2*). The subsequent chapter addresses open innovation platforms and derives a process for designing an OIP from socio-technical perspective (*chapter 3*). Subsequently, the OIP lifecycle model is set out in order to describe activities in each phase of it (*chapter 4*). Additionally, mechanisms to overcome major challenges in the lifecycle of OIPs are analyzed (*chapter 5*). The key findings of this part are summarized in *chapter 6*.

Part V - Conclusion closes the work with an overall summary of the contributions and findings of this thesis (*chapter 1*). Furthermore, based on the results of the present study, managerial implications (*chapter 2*) as well as further directions for research (*chapter 3)* are derived. The thesis closes with a note on how parts of this thesis were communicated to the research community (*chapter 4*).

Part II

Foundations

1 Open innovation platforms

Part I identified three facets of OIP lifecycle management. The following *Part II* sheds light on these three facets: open innovation platforms (*chapter 1*), open innovation intermediaries (*chapter 2*) and lifecycle models for OIPs (*chapter 3*). It closes with a short summary of the major findings (*chapter 4*) [63].

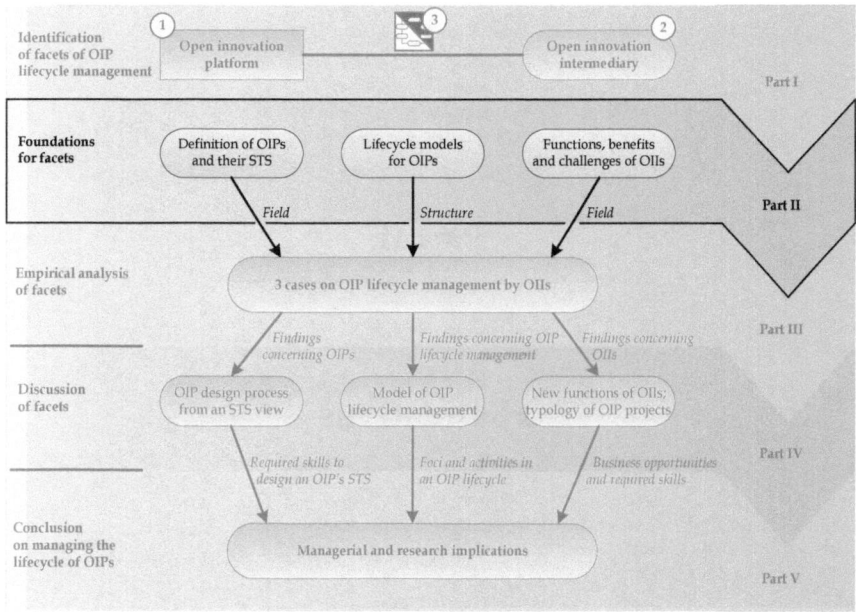

Figure 4: Current progress in the research design

This first chapter aims at defining and characterizing open innovation platforms and their socio-technical system. It, thus, sets the foundations to derive a design process for OIPs from a socio-technical perspective in Part IV. The combination of the terms 'open innovation' and 'platform' indicates that OIPs refer to IT-based platforms that are used to conduct open innovation. Therefore, an open innovation platform (OIP) provides a virtual environment to facilitate the interaction among *organizers* and

---

[63] See Figure 4.

innovators. The former seek innovative solutions to a problem[64], whereas the latter help to solve the problem. This chapter outlines how OIPs support the connection between organizers and innovators.

Firstly, relevant streams from the literature concerning OIPs are summarized (*section 1.1*). Secondly, a definition of OIPs is derived and terms regarding OIPs are defined (*section 1.2*). Having cleared these basics, details on the IT-based tools for open innovation that can be implemented on an OIP are introduced (*section 1.3*). Finally, the socio-technical system of an OIP is described (*section 1.4*).

1.1   Relevant research streams for this chapter

Two overlapping streams of research are currently present in the literature on open innovation[65]: The first stream follows Henry Chesbrough's[66] perception who sees open innovation as the opposite of closed innovation. The focus of this stream lies on *opportunities* and *implications for organizations*, such as the following: The acquisition of external expertise[67], dealing with intellectual property rights[68], establishing a strategic advantage[69] and ways to overcome problems arising from the not-invented-here (NIH) syndrome[70]. Thus, new business models for outside-in, inside-out and coupled process innovations are analyzed[71]. The second stream follows Eric von Hippel[72] who takes a *user and tool perspective*[73]. Users collaborate[74] in order to develop innovations in a virtual environment[75]. Topics of this stream include the user's motivation to voluntarily contribute to an organizer's problem[76], the identification[77] and attraction[78]

---

[64] For more details on organizers of OIPs see section 1.2.
[65] Moeslein and Neyer (2009).
[66] Chesbrough (2003).
[67] Chesbrough (2003); Dahlander and Wallin (2006); Neyer, Bullinger and Moeslein (2009).
[68] Chesbrough (2006a); Huizingh (2011); Terwiesch and Xu (2008).
[69] Dahlander and Gann (2010).
[70] Gassmann (2006); Piller and Walcher (2006).
[71] Chesbrough (2006b); Gassmann and Enkel (2004).
[72] Von Hippel (2005).
[73] See Bogers, Afuah and Bastian (2010).
[74] Greer and Lei (2012).
[75] Von Hippel and Katz (2002).
[76] Fueller (2006); Harhoff (2003).
[77] Belz and Baumbach (2010); von Hippel, Franke and Pruegl (2009); Morrison, Roberts and Midgley (2004).
[78] Adamczyk et al. (2010); Bullinger and Moeslein (2011).

of appropriate innovators and ways to design the tools for open innovation that are in this thesis considered OIPs[79].

The following analysis integrates both views on open innovation. The Chesbroughian view is reflected by organizations who use OIPs to open up their innovation process. Von Hippel's view is incorporated by the users of an OIP, i.e. the innovators and the focus on IT-based tools for open innovation.

The last section of this chapter introduces OIPs as socio-technical system. Therefore, it will refer to foundations that concern the socio-technical systems theory[80] stemming from former work system theory in organizational sciences[81]. Also, more recent work from the field of information systems drawing on the socio-technical systems theory is used to describe the socio-technical system of OIPs[82]. This section starts with a definition of OIPs.

1.2   Definition of open innovation platforms

Chesbrough[83] notes the use of IT-based tools for open innovation, namely technology marketplaces, in his first book on open innovation. Thus, the use of IT-based tools is a part of open innovation in Chesbrough's perception from the very beginning. Tools for IT-based open innovation[84] share a common topical ('open innovation') and technical ('platform') foundation, which is explained below.

From a *topical perspective* OIPs comprise IT-based tools that facilitate open innovation. They thus allow integrating innovators into the innovation process of an organizer[85] by supporting the search for, the selection of and/or the implementation of innovations[86].

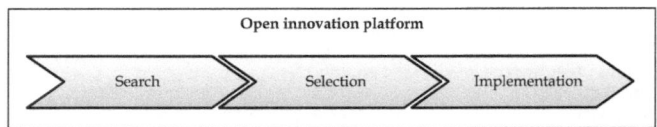

Figure 5: Steps of an innovation process

---

[79]   Bullinger and Moeslein (2011); Leimeister, Huber, Bretschneider and Krcmar (2009); Nambisan (2002).
[80]   Trist and Bamforth (1951).
[81]   Trist, Higgin, Murray and Pollock (1963); Trist (1981).
[82]   For instance Bostrom and Heinen (1977); Krcmar (2010); Sommerville (2011).
[83]   Chesbrough (2006b).
[84]   For a list of IT-based tools for open innovation see section 1.3.
[85]   An organizer of an OIP might for instance be a company. Details are given below.
[86]   See Tidd and Bessant (2009) and Figure 5.

During the first process step, the *search*, either existing suggestions are gathered or new suggestions are generated. It is the aim to compile a pool of creative suggestions. In the case of open innovation, both company internal (e.g. employees) and external sources (e.g. consumers, suppliers, experts) contribute to this process. The elaboration level of suggestions varies between short descriptions and completely implemented solutions[87]. However, usually ideas or concepts are searched for.

During the second step of the innovation process, the *selection*, good suggestions have to be identified and separated from the less suitable ones. This selection can be done either by experts or by a community[88]. Typical approaches of community evaluation are (1) the vote, where consent or refusal establish an order of priority; (2) forming an average, where multi-stage (Likert) scales are used to define threshold values, for example; (3) the accordance, where e.g. comparisons or comments are consulted in pairs to establish orders of priority[89].

During the third step, the *implementation*, selected suggestions are turned into products or services and introduced into the market[90]. The step from a suggestion to a completed product or service is costly, that is why in practice companies often create prototypes before the complete implementation[91]. Within the context of open innovation it is shown that these prototype implementations are partly created by the innovators[92]. In this way, the biggest amount possible of implicit knowledge is transferred to the company or to the community and the step from initial suggestions to final implementations is easier.

Concerning the definition of innovators, this thesis follows the perception of Neyer, Bullinger and Moeslein[93]. Accordingly, the term innovator comprises internal (e.g. core research and development employees), peripheral (e.g. other employees of an organization) and external innovators (e.g. customers, suppliers, experts). OIPs depend on the voluntary participation of these innovators[94].

---

[87] Adamczyk, Bullinger and Moeslein (2012); Hallerstede et al. (2010).
[88] Bullinger, Neyer, Rass and Moeslein (2010).
[89] Haller (2012); Riedl, Blohm, Leimeister and Krcmar (2010).
[90] Tidd and Bessant (2009).
[91] Von Hippel and Katz (2002).
[92] Bullinger, Hoffmann and Leimeister (2011); von Hippel and Katz (2002).
[93] Neyer et al. (2009).
[94] Bateman, Gray and Butler (2010); Harhoff (2003); Moon and Sproull (2008).

From a *technical perspective*, OIPs are virtual environments, which serve as infrastructure for the exchange of information[95]. These environments particularly integrate IT-based tools for open innovation[96]. Most of them are realized as web-based platforms[97], which means that they offer a web-based interface as a mean of interaction for the users, e.g. the innovators, with the OIP[98]. Those platforms are defined, virtual environments that provide digital services[99]. They allow for interaction without time- or location-based constraints[100]. In order to provide an OIP, hardware[101], software[102] and services[103] are required. From this topical and technical perspective

> *an open innovation platform (OIP) is defined as a virtual environment that offers digital services, with the aim to allow the creation of innovations by facilitating time- and location-independent, voluntary interaction of innovators.*

According to this definition, OIPs are a sub-class of interaction platforms as defined by Ihlenburg[104]. In order to characterize OIPs further and define terms used in the context of OIPs, design elements to describe OIPs are introduced in the following.

Literature offers a magnitude of design elements to describe different tools for IT-based open innovation[105]. The following paragraphs will draw on this pool of design elements and derive a subset of design elements to describe OIPs, namely a set of design elements all tools of IT-based open innovation share[106]. The design elements and their attributes are summarized in Figure 6.

An open innovation platform is initiated by an *organizer* who is interested in the innovation outcome[107]. The organizer can be a company, a public or non-profit organization or an individual. Usually, the organizer dedicates the OIP to a *specific*

---

[95] Zerdick et al. (2001).
[96] IT-based tools for open innovation are elaborated on in section 1.3.
[97] See Diener and Piller (2010).
[98] Komus and Wauch (2008).
[99] In this case interfaces.
[100] Ihlenburg (2011).
[101] For instance servers.
[102] For instance the source code of the platform or a webserver.
[103] For instance the internet connection in a data center.
[104] Ihlenburg (2011).
[105] See for instance Adamczyk et al. (2012); Haller, Neyer and Bullinger (2009).
[106] The selection of design elements is based on the review of research and practice by Bullinger and Moeslein (2011).
[107] Ebner, Leimeister and Krcmar (2010); Klein and Lechner (2009); McWilliam (2000); Smith, Banzaert and Susnowitz (2003).

topic; details of which vary extensively[108]. An innovation community dedicated to a brand for instance has rather low topic specificity as innovators can contribute on any topic concerning the brand. On the other hand, looking at a configurator, the specificity of the topic might be high, as the configurator might limit the solution space[109]. The topic indicates the desired *degree of elaboration* of contributions[110]. They range from ideas, sketches and concepts to virtual and physical prototypes or finished solutions. In addition, different degrees of elaboration of contributions might be accepted at the same time or they could evolve over time[111].

Design element: definition	Attributes						
Organizer: entity initiating an OIP	Company	Public organization		Non-profit		Individual	
Topic specificity: solution space for contributions	Low (open topic)		Defined			High (specific topic)	
Degree of elaboration: required level of detail for contributions	Idea	Sketch	Concept	Prototype	Solution	Mixed	Evolving
Target group: description of participants	Specified			Unspecified			
Participation as: number of persons forming one entity of participant	Individual		Team		Both		
Runtime: runtime of OIP	Very short		Short		Long		Very long
Tool: tool provided for interaction	Innovation contest	Innovation community	Innovation market place	Innovation toolkit		Mixed	
Motivation: incentives for participation	Monetary		Non-monetary			Mixed	
Evaluation: method to determine ranking of contributions	Jury evaluation	Peer review		Self-assessment		Mixed	

Figure 6: Design elements characterizing open innovation platforms[112]

By defining the topic, the organizer also indicates the intended *target group*[113] and the mode of *participation*[114]. The target group might be either specific, e.g. a particular consumer segment or region, or unspecific, e.g. a broad audience. Innovators from the target group might contribute as individuals, in teams or both modes might be accepted. Each OIP is online for a limited period of time; during this *runtime*, submitting contributions is possible[115]. Periods range from days (very short runtime)

---

[108] Ebner et al. (2010).
[109] See Von Hippel and Katz (2002).
[110] Ebner et al. (2010); von Hippel and Katz (2002); Klein and Lechner (2009); Smith et al. (2003); Soll (2006).
[111] Adamczyk et al. (2012); Haller (2012).
[112] Building on Bullinger and Moeslein (2011).
[113] Brabham (2010); Bullinger, Haller and Moeslein (2009); Carvalho (2009); Hallerstede et al. (2010).
[114] Boudreau, Lacetera and Lakhani (2008); Carvalho (2009); Smith et al. (2003).
[115] Bullinger et al. (2009); Jeppesen (2005); Walcher (2007).

over weeks (short runtime) to multiple months (long runtime) or even years, which
can be interpreted as ongoing OIPs (very long runtime)[116]. While innovation contests
for instance tend to have very short or short runtimes[117], innovation communities and
innovation marketplaces are usually installed for a long to very long runtime. The
OIP can offer a single *tool* for IT-based open innovation or a combination of tools that
facilitate interaction[118]. These tools are used to structure the OIP's innovation process.
Innovation contests, innovation communities, innovation market places, innovation
toolkits or a combination of these are classes of tools that can be integrated in an OIP.
To foster participation a motivational system is established and adjusted to the target
group's needs. *Motivation* can be induced via monetary or non-monetary
incentives[119]. While the first includes awards and prizes, the latter refers to motivators
like community feedback, the establishment of a reputation among relevant peers, or
self-realization. All of these factors can be supported[120]. An appropriate motivation
mechanism depends for instance on the topic of the OIP[121]. Once contributions are
made, different groups can carry out their *evaluation*[122]. On some OIPs, a jury of
experts evaluates contributions, but also peer reviews and self-assessments are used.
Furthermore, a combination of methods, e.g. a peer-review followed by a jury
evaluation might apply. The following section sets out the IT-based tools for open
innovation an OIP can implement to facilitate interaction.

1.3   IT-based tools for open innovation constituting an OIP

There is research on different IT-based tools for open innovation[123]. However, there is
no compelling review or taxonomy that structures available tools. Moeslein and
Neyer [124] compiled a list of tools for IT-based open innovation based on their
experience. They distinguish between five classes, namely *innovation contest,*
innovation communities, innovation market places, innovation toolkits and *innovation*

---

[116] As requested by Bullinger and Moeslein (2011), the runtime's attributes are adjusted to the
    increasing runtime of OIPs. Originally, the runtime was defined as some hours (very short
    runtime), weeks (short runtime), months (long runtime) and years (very long runtime). These
    ranges are adjusted to the ones presented above.
[117] Bullinger and Moeslein (2011).
[118] See section 1.3 and Ihlenburg (2011).
[119] Boudreau et al. (2008); Bullinger et al. (2009); Ernst (2004); Fueller (2006); Harhoff (2003); Ogawa
    and Piller (2006); Piller and Walcher (2006).
[120] Brabham (2009); Harhoff (2003); Piller and Walcher (2006).
[121] Hallerstede and Bullinger (2010); Terwiesch and Xu (2008).
[122] Carvalho (2009); Ebner et al. (2010); Haller (2012); Klein and Lechner (2009); Nambisan (2002).
[123] Gassmann et al. (2010); Hrastinski et al. (2010).
[124] Moeslein and Neyer (2009).

technologies. This list claims neither mutual exclusiveness nor collective exhaustiveness, but serves as a starting point for stereotypes of tools for IT-based open innovation[125]. The five classes can be seen as tools, which are combinable if necessary, in order to design an open innovation platform. The following subsections briefly introduce the five classes of tools for IT-based open innovation.

1.3.1 Innovation contests

Innovation contests are web-based competitions of innovators who use their skills, experiences and creativity to provide a solution for a particular challenge in order to generate innovations[126]. They use web 2.0[127] mechanisms to transfer contributions of innovators to an organizer[128]. Innovators work by themselves or in groups to develop suggestions to a particular problem[129], which are submitted via a platform to an organizer[130]. The evaluation of suggestions might be carried out by peers, i.e. the innovators, or by a jury of experts[131]. The winning contributor(s) gain(s) a reward[132]. Thus, the main focus in innovation contests is competitively solving a challenge[133].

The major strength of innovation contests lies in the search for innovative ideas. However, the selection and implementation phase of the innovation process can be supported as well. Examples for innovation contests include the Swirl Smell Fighter Contest to generate new ideas against domestic smell[134], the Google Lunar X-Prize to fly to the moon[135], the Virgin Earth Challenge to stabilize the Earth's climate system[136] and Stilsicher:unterwegs to develop new walking frames[137]. A list with more examples is accessible on innovationresearch.de.

---

[125] A comprehensive assessment and a taxonomy of IT-based tools for open innovation is not the focus of this thesis. This gap in research is considered a future research need.
[126] See Bullinger and Moeslein (2011).
[127] O'Reilly (2005).
[128] Adamczyk et al. (2012); Bullinger et al. (2010); Jain (2010).
[129] Boudreau et al. (2008); Bullinger et al. (2010); Carvalho (2009); Smith et al. (2003).
[130] Boudreau et al. (2008); Brabham (2009); Piller and Walcher (2006).
[131] Carvalho (2009); Ebner et al. (2010). See also section 1.2.
[132] Haller, Bullinger and Moeslein (2011).
[133] Terwiesch and Xu (2008).
[134] www.smellfighters.com; retrieved July 22, 2012.
[135] www.googlelunarxprize.org; retrieved July 22, 2012.
[136] www.virgin.com/subsites/virginearth; retrieved July 22, 2012.
[137] www.stilsicher-unterwegs.de; retrieved July 22, 2012.

1.3.2   Innovation communities

Opposing to the competitive setup in innovation contests, innovation communities build upon a collaborative generation of innovations[138]. While communities differ in structure, topical focus and the extent of social ties[139], they are mainly based on shared enthusiasm and knowledge concerning a particular domain[140]. On an innovation community platform, community members for instance discuss new ideas for products and services or the question how to improve them[141]. The willingness to contribute thus depends on the particular characteristics of the community, like language, netiquette, norms and the individual motivation of the community members[142]. Communities can implement characteristics of innovation contests by hosting a challenge within a community[143]. In this case, challenges tend to offer low monetary or non-monetary rewards to avoid undermining the collaborative community spirit[144].

Innovation communities can help to develop and evaluate innovative ideas and thus mainly support searching and selecting in the innovation process. Anyway, community members might opt to implement prototypes or complete solutions, which supports the implementation phase. Examples of innovation communities include the Dell IdeaStorm to improve Dell's products[145], the The Forge community by Local Motors to develop the innovations in the automotive sector[146], the Open Architecture Network to improve living conditions through innovative design[147] and Gemeinsam fuer die Seltenen to identify problems of and corresponding solutions for patients with rare diseases[148]. A list with more examples is accessible on open-innovation-projects.org.

---

[138] Hutter et al. (2011).
[139] Armstrong and Hagel III (1996); Fueller et al. (2006); Lynn, Aram and Reddy (1997); Muniz Jr. and O'Guinn (2001); Rheingold (2000).
[140] Kozinets (1999).
[141] Fueller, Matzler and Hoppe (2008); von Hippel (2005).
[142] Armstrong and Hagel III (1996); Bateman et al. (2010); Butler, Sproull, Kiesler and Kraut (2007); von Hippel (2005); Williams (2000).
[143] Bullinger et al. (2010); Hutter et al. (2011).
[144] Hutter et al. (2011).
[145] www.ideastorm.com; retrieved July 22, 2012.
[146] forge.localmotors.com; retrieved November 19, 2012.
[147] www.openarchitecturenetwork.org; retrieved July 22, 2012.
[148] www.gemeinsamselten.de; retrieved July 22, 2012.

1.3.3   Innovation market places

Innovation marketplaces either provide the opportunity to post problems to which innovators can suggest solutions or vice versa: Innovators post their solution to a problem in order to find someone who needs it[149]. An open innovation intermediary that runs the innovation market place usually maintains a community to solve the problems of organizers[150]. The organizer grants the innovator a payment of which the OII retains a commission. Innovation market places generate innovations by providing a mean to combine existing knowledge or approaches with new areas of application. A project on an innovation marketplace where an organizer asks for a solution can be seen as an innovation contest. Thus, an innovation market place offers a default design for multiple concurrent innovation contests.

Innovation marketplaces help to find new ideas, concepts or finished solutions. They thus support all phases of the innovation process. Among innovation market places are for example the following: In Yet2[151] and Innocentive[152] mostly solutions to highly complex problems are sought from or provided by professionals[153]. In Battle of Concepts projects seek for concepts by tinkerers[154]. Brainfloor conducts brainstorming projects with a general audience[155]. See Diener and Piller[156] for more examples of innovation marketplaces.

1.3.4   Innovation toolkits

Innovation toolkits provide a virtual environment with a limited solution space in which innovators can create innovative solutions based on a defined process[157]. By providing a limited solution space, implementability of the results can be ensured, as innovators do not need any particular knowledge in the field. There are three basic types of innovation toolkits: Innovation toolkits for user innovation, innovation toolkits for user co-design and innovation toolkits for idea transfer[158]. On the one hand, innovation toolkits limit the innovators creativity by providing boundaries, i.e.

---

[149] Bakos (1997); Lichtenthaler and Ernst (2008).
[150] Diener and Piller (2010).
[151] www.yet2.com; retrieved July 22, 2012.
[152] www.innocentive.com; retrieved July 22, 2012.
[153] Bishop (2009). The terms professionals, tinkerers and general audience refer to types of innovators as set out in Hallerstede et al. (2010).
[154] www.battleofconcepts.nl; retrieved July 22, 2012.
[155] www.brainfloor.com; retrieved July 22, 2012.
[156] Diener and Piller (2010).
[157] Von Hippel and Katz (2002).
[158] Reichwald and Piller (2009).

the limited solution space, but on the other hand, they foster creativity by giving impulses towards a desired direction of contributions[159].

Innovation toolkits help to transfer tacit knowledge of innovators to the organizer of the innovation toolkit by providing prototype-like contributions[160]. Thus, the search for ideas and the implementation of them is the main focus of innovation toolkits. Examples of innovation toolkits include miadidas to design customized shoes[161], Lego Cuusoo to create and share custom Lego builds[162], open source libraries like the Apache security software[163] being in use by approximately 70% of today's websites[164] and prototype packages applied by unserAller to create physical prototypes[165].

1.3.5   Innovation technologies

Innovation technologies support the implementation of ideas. Therefore, they offer a platform or interface to transmit custom designs[166] to, or receive input from, rapid prototyping devices like 3D printers, 3D scanners or laser cutters. The innovator is thus capable to generate physical prototypes from its custom design. These prototypes thereby serve multiple purposes in the innovation process[167]. They are mostly used to elicit requirements for a product[168], to improve the communication among innovators by serving as a basis for discussions[169] or to evaluate intermediary results[170]. A combination of these benefits is also possible[171]. These implementations go beyond those from innovation toolkits, as they bridge the gap from virtual proto-types created with innovation toolkits to physical ones using innovation technologies.

Innovation technologies enable innovators to implement their ideas. Thus, the focus of this tool is the implementation phase, but as outlined above, innovation technologies also support the search and selection phases. Examples for innovation technologies include Ponoko, who offers the opportunity to create and distribute

---

[159] Piller and Walcher (2006); Toubia (2006).
[160] Von Hippel and Katz (2002).
[161] www.miadidas.com; retrieved July 22, 2012.
[162] lego.cuusoo.com; retrieved November 19, 2012.
[163] httpd.apache.org; retrieved July 22, 2012.
[164] Franke and Von Hippel (2003); Netcraft (2012).
[165] www.unseraller.de; retrieved July 22, 2012.
[166] Designs are transferred using CAD (computer-aided design) data.
[167] Bullinger et al. (2011).
[168] Cao and Ramesh (2008).
[169] Dix, Finlay, Abowd and Beale (2004).
[170] Davis and Venkatesh (2004).
[171] Davis (1992); Kordon and Luqi (2002).

custom designs using a configurator[172], eMachineShop to produce metal prototypes from CAD data[173] and ShapeWays to produce prototypes using different materials[174].

1.3.6 Scope of this thesis

Innovation technologies provide a specific technical interface to a physical device. Thus, they focus on hardware and do not provide a virtual environment to the users, such as a platform. In addition to that, lifecycle management for hardware differs from lifecycle management for software[175]. The latter one is, however, the emphasis of this thesis. Due to these reasons, innovation technologies are not in the focus of this study and not subsumed under the term OIP. They are rather captured indirectly, as they tend to apply innovation communities and innovation toolkits as an interface, i.e. the platform, for innovators. Thus, in the perception of this thesis, OIPs refer to virtual environments that incorporate *innovation contests, innovation market places, innovation communities, innovation toolkits* or a *combination thereof.*

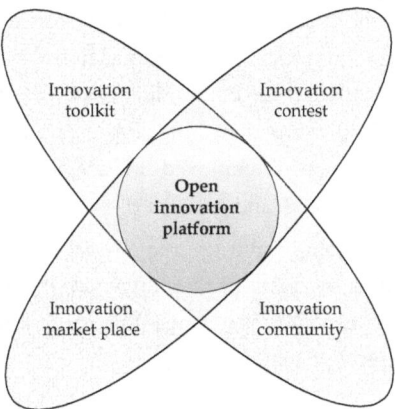

Figure 7: IT-based tools for open innovation that constitute an OIP

OIPs and their design elements contain both, social[176] as well as technical[177] dimensions. Thus, when managing the lifecycle of an OIP, social *and* technical aspects

---

[172] www.ponoko.com; retrieved July 22, 2012.

[173] www.emachineshop.com; retrieved July 22, 2012.

[174] www.shapeways.com; retrieved July 22, 2012.

[175] Lee et al. (2009).

[176] For instance the target group.

[177] For instance a tool's implementation.

have to be considered as already proposed by socio-technical systems thinkers in former studies in a multitude of settings[178]. Therefore, the following section will investigate OIPs from a socio-technical systems perspective.

1.4   Open innovation platforms from a socio-technical systems perspective

The socio-technical systems theory origins from observations in the UK coal mining industry in the early 1950s. Trist and Bamforth [179] noticed that a common technological innovation, i.e. the application of the longwall method, led to varying output of different coal mines. In some mines, the technological innovation came along with a new "one man – one job"[180] policy, which promised the workers more autonomy and a more flexible work assignment than the rigid work group organization used before. The researchers found out that applying this policy resulted in a higher output of the coal mine even though applying the same mining technology. They concluded that the applied technology alone did not suffice to explain productivity of a coal mine and highlighted the influence of the social structure established in the different mines[181]. Further research led to the conclusion that considering organizations as an "'open socio-technical system' helps to provide a more realistic picture of how they are both influenced by and able to act back on their environment"[182], which further broadened the view on a socio-technical system. Since then, the STS theory has been employed in a wide variety of settings and disciplines[183] including IS research and practice[184].

A socio-technical system is split up into a *social* and a *technical subsystem* that turn *input* into *output* in an *environment* that influences the STS[185]. This environment refers to legislation, culture, norms, availability of resources and other factors influencing a socio-technical system[186].

---

[178] See for instance Haller (2012); Neyer et al. (2009); Pasmore et al. (1982); Trist and Bamforth (1951).
[179] Trist and Bamforth (1951).
[180] Trist et al. (1963), p. 13.
[181] Trist and Bamforth (1951).
[182] Emery and Trist (1960), p. 94.
[183] Pasmore et al. (1982); Trist (1981).
[184] Bostrom and Heinen (1977); Krcmar (2010); Sommerville (2011).
[185] See Figure 8.
[186] Cummings and Srivastva (1977).

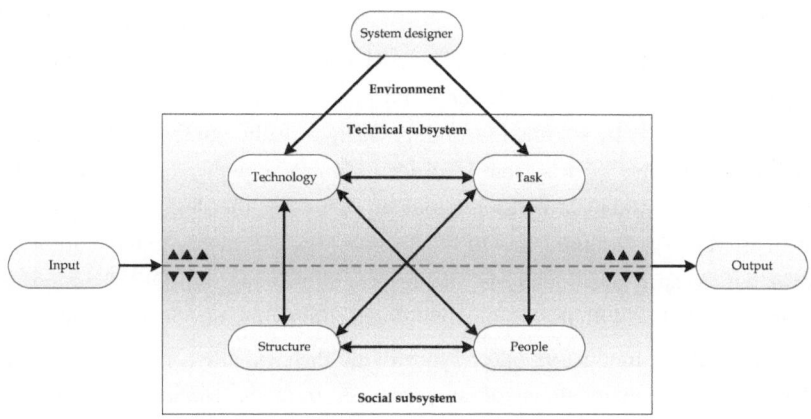

Figure 8: The socio-technical systems theory[187]

The *technical subsystem* concerns the *task* of, for instance, an organization and the *technology* to accomplish the tasks. Consequently, a technology does not necessarily refer to a technical implementation, but also to an artifact that provides structure like a tool or process. The technical subsystem shapes the challenge, variety, feedback, control, integration and decision-making of people working with it[188].

The *social subsystem* contains the attributes of *people*, the relationships among them, as well as the *structures* people work in. This includes for instance skills, attitudes, reward systems and authority structures. In a broad sense, the social subsystem comprises all factors of why humans decide to work in an organization including their attitudes towards and their expectations of the organization. Thus, identifying people's needs and meeting those needs in the technical subsystem is the best way to direct efforts towards a desired output[189].

Though the social and the technical subsystem are basically independent, they influence each other[190]. Thus, considering *either* the social *or* the technical subsystem does not suffice to design a work system. The two systems rather call for joint design and optimization while considering the interactions with the environment[191].

[187] Adapted from Bostrom and Heinen (1977).
[188] Bostrom and Heinen (1977); Geels (2004); Pasmore et al. (1982).
[189] Bostrom and Heinen (1977); Geels (2004); Pasmore et al. (1982).
[190] Pasmore et al. (1982).
[191] Cherns (1976); Cummings and Srivastva (1977); Emery and Trist (1960); Fox (1995).

Bostrom and Heinen[192] adapt the socio-technical systems theory to an IS context and argue that a designer of a socio-technical system can only influence technologies and tasks directly[193]. Structures and people of the social subsystem are only influenced indirectly by so called secondary changes. Although the authors argue for changing an existing work system in order to improve quality or task accomplishment[194], the socio-technical considerations are valid for the design of a new work system as well, e.g. designing an OIP[195]. Applying the STS theory to the field of OIPs allows for a structured analysis of factors influencing an OIP's design and management. In line with this argumentation, the following introduces an OIP's STS.

An OIP ask innovators *(people)* to provide their expertise and ideas *(input)* in order to develop innovations for an organizer *(task and output)* using provided functionality *(technology)* in a given relational and motivational context *(structure)*. The social and technical subsystem is influenced by external conditions that concern the OIP *(environment)*.

The *environment* includes factors that influence the OIP like legislation, corporate and country culture, an organizers' corporate guidelines and availability of resources concerning design and management of the OIP. The OIP is designed to facilitate the *task* of innovating for the organizer of the OIP. The task may vary in specificity (open vs. detailed) and ask for a certain degree of elaboration of the output (idea to functional product)[196]. *Technology* is provided to accomplish the task. The technology defines the OIP's processes, architecture, functionalities and graphical design. Innovators can for instance only submit suggestions that are supported by the technical subsystem. In the context of software development, the technical subsystem also includes implementations of the technologies. The *structure* comprises process governance, the OIP's organization, culture and rules. This includes in particular the reward systems, the intentions of the graphical design and the influence of community managers and other actors on the processes and outputs. Also, the opportunities people have to interact are part of the structure. *People* in the context of an OIP encompass multiple parties, including innovators, organizers, community managers as well as OIP lifecycle managers[197]. The innovators' characteristics are of

---

[192] Bostrom and Heinen (1977). See Figure 8.
[193] See *system designer* in Figure 8.
[194] Bostrom and Heinen (1977).
[195] See for instance Bryl, Giorgini and Mylopoulos (2009); Clegg (2000).
[196] Bullinger and Moeslein (2011). See also section 1.2.
[197] For details on OIP lifecycle managers see chapter 2.

particular interest, since relevant people have to be attracted for a task. These people volunteer to *input* their time, expertise and ideas in order to generate innovations for an organizer. Table 1 summarizes the components of an OIP's socio-technical system.

Table 1: Components of an OIP's socio-technical system

STS component	Interpretation in the context of OIPs
Environment	Legislation, corporate and country culture, an organizers' (e.g. corporate) guidelines and availability of resources
Input	Innovators' ideas, knowledge and creativity
Task	Generation of innovations for an organizer
Technology	Functionalities, processes, rules, implementations and graphical design
People	Innovators and community managers that work on the OIP
Structure	Organization and governance of the OIP's processes including motivational aspects
Output	Innovations for the organizer

The socio-technical systems theory was originally designed for work systems in which its people are members of an organization. This setting provided a given structure and a known set of people. In the context of OIPs, innovators are not organized in an organization, but in a loose network of voluntary participants[198]. Thus, there are no compulsive organizational structures or predefined people who influence the system. Two implications result from this situation:

Firstly, as the *technology* is the main point of contact and mean of interaction for the innovators, it has to communicate the *task*, convey the intended *structure* in terms of motivation and relationships among innovators and it has to attract relevant *people*. To sum it up, the technical subsystem has to inform and guide the innovators that use the OIP. One might picture an innovator visiting the OIP with the aim to participate. The OIP, i.e. the technical subsystem, has to give them advise on what to do and how to do it, as they do not have any information on the given task or any structure that could guide them.

Secondly, the designer of an OIP can influence the social subsystem only *indirectly*. Anyhow, the OIP designer can *directly* influence the technical subsystem[199]. Thus, as the components of a STS are interdependent, the designer of an OIP can, and

---

[198] Adamczyk et al. (2010); Borst (2010); Pittaway et al. (2004).
[199] Bostrom and Heinen (1977).

has to, shape the intended *social* subsystem indirectly by designing a well mapped-out *technical* subsystem. This includes technology *and* task design. Accordingly, the OIP designer has to estimate the impact of a manifestation in the technical subsystem on the social subsystem. One might think of an innovator visiting the OIP. A serious businessman would probably not be attracted by a colorful and playful graphical design, though a hobbyist possibly would. By this technical manifestation (i.e. the graphical design) the OIP designer can attract a certain group of people with certain characteristics and thus influence the social subsystem (i.e. the characteristics of the participants). Based on the present chapter, which has defined and characterized OIPs, the following chapter addresses the lifecycle managers of OIPs, namely open innovation intermediaries.

2 Open innovation intermediaries

Long before open innovation and even innovation was subject to research, innovation intermediaries arose due to the fact that there has always been a gap between organizers[200] that seek solutions to an innovation problem and innovators that can provide a solution to an organizer's problem. An early example is the 16th century wool industry in which traders ('middlemen') functioned as informal disseminators of knowledge of technical improvements for wool processing[201]. Following this early and generic[202] picture of innovation intermediaries, an

> *innovation intermediary is defined as an organization that bridges the gap between organizers that seek solutions to an innovation problem and innovators that can provide a solution to an organizer's problem[203].*

The present chapter aims at introducing innovation intermediaries with respect to the particular circumstances in open innovation. Innovation intermediaries that use open innovation platforms[204] to integrate innovators in their client's (i.e. an organizer's) innovation process are called open innovation intermediaries[205]. Accordingly, an

> *open innovation intermediary (OII) is defined as an organization that uses open innovation platforms to bridge the gap between organizers that seek solutions to an innovation problem and innovators that can provide a solution to an organizer's problem.*

Figure 9 illustrates the definition of innovation intermediary and open innovation intermediary[206]. These foundations help to characterize OIIs and derive a typology of

---

[200] An organizer can be either a company, a public or non-profit organization or an individual. See section 1.2.

[201] Farnie (1979); Hill (1967); Howells (2006); Smith (2002).

[202] Dalziel (2010).

[203] Cf. Dalziel (2010); Datta (2007); Hargadon and Sutton (1997); Howells (2006); Sapsed, Grantham and DeFillippi (2007); Sieg et al. (2010); Verona et al. (2006). This generic definition of innovation intermediaries includes a diversity of terms for innovation intermediaries including bridge builders (Tidd and Bessant, 2009), intermediary firms (Stankiewicz, 1995), intermediaries (Callon, 1994; Watkins and Horley, 1986), technology brokers (Provan and Human, 1999), boundary organizations (Cash, 2001; Guston, 1999) and virtual knowledge brokers (Verona et al., 2006).

[204] See chapter 1.

[205] See also Hossain (2012).

[206] According to this definition, an OII represents a type of innovation intermediary, that externally sources ideas, as proposed by Lopez-Vega and Vanhaverbeke (2009).

OIP projects in Part IV. Based on this, organizers of an OIP will be able to tell which OII is suitable to run a particular type of OIP project for them.

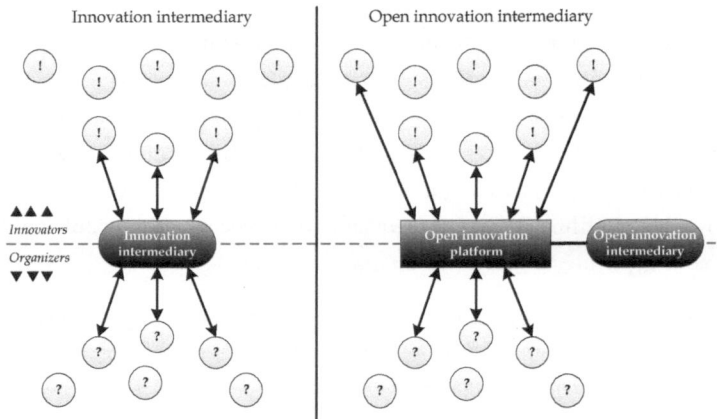

Figure 9: Innovation intermediary versus open innovation intermediary bridging the gap between organizers and innovators

The sections in this chapter present the investigated research streams that address open innovation intermediaries (*section 2.1*), the question which functions open innovation intermediaries can offer (*section 2.2*), reasons why organizations should rely on open innovation intermediaries (*section 2.3*), and also the question which challenges arise for open innovation intermediaries when working in a virtual environment (*section 2.4*).

2.1   Relevant research streams for this chapter

Innovation intermediaries are known from different, partly overlapping literature streams, which all set a distinct emphasis on the role of innovation intermediaries[207]. The first stream is the *diffusion and technology transfer* literature where intermediaries improve the speed of diffusion and new product uptake by supporting decision making[208], by creating and supporting relationships among organizations[209], by formalizing informal collaborations, and by providing negotiation and contractual skills[210]. The second stream is the *innovation management* literature that focuses on

---

[207] Howells (2006); Lopez-Vega and Vanhaverbeke (2009).
[208] Haegerstrand (1952); Mantel and Rosegger (1987); Rogers (2003).
[209] Watkins and Horley (1986).
[210] Seaton and Cordey-Hayes (1993); Shohet and Prevezer (1996).

innovation intermediaries' activities. There, the key function of innovation intermediaries is to initiate and support the technology transfer process across people,
organizations and industries[211]. Besides connecting clients, innovation intermediaries
store and transform knowledge for them[212]. Thirdly, the *systems and networks*
literature takes a broader look on innovation intermediaries and their influence on the
overall innovation system and policy. They link players in the market[213] or influence
the system on a strategic level[214]. The last contributing stream of research origins from
service innovation. In this field of knowledge intensive business services innovation
intermediaries build up a close and continuous interaction with their clients. This
way, they support innovative change and act as bridges for innovation[215].

Howells[216] merges the findings from these diverse literature streams into his
review on innovation intermediaries resulting in the functions of innovation
intermediaries introduced below. I update his findings by integrating more recent
research results stemming from open innovation literature[217].

2.2   Functions of open innovation intermediaries

Innovation intermediaries, in their traditional notion, and thus open innovation intermediaries as particular representatives of innovation intermediaries, can fulfill
multiple functions for their clients[218]. Lopez-Vega and Vanhaverbeke[219] structure the
functions of open innovation intermediaries along the three dimensions *connection,*
collaboration as well as *support and technological services*, as shown in Table 2.

Firstly, open innovation intermediaries *connect* their clients, i.e. organizers that
seek solutions to an innovation problem, and innovators, i.e. individuals that can
provide a solution to an organizer's problem, on different levels[220]. These levels
include linking organizations to customers[221], non-customers[222], peers[223], initiatives[224],

---

[211] Hargadon and Sutton (1997); McEvily and Zaheer (1999).
[212] Hargadon and Sutton (1997).
[213] Cash (2001); Lynn, Mohan Reddy and Aram (1996); Stankiewicz (1995).
[214] Braun and Guston (2003); Braun (1993); Callon (1980), (1994); Cash (2001); Guston (1996), (1999);
      Kelly (2003); van der Meulen and Rip (1998).
[215] Bessant and Rush (1995); Czarnitzki and Spielkamp (2000); Howells (1999); Wood (2002).
[216] Howells (2006).
[217] See for instance Diener and Piller (2010); Lopez-Vega and Vanhaverbeke (2009).
[218] Howells (2006); Lopez-Vega and Vanhaverbeke (2009).
[219] Lopez-Vega and Vanhaverbeke (2009).
[220] Klerkx and Leeuwis (2009); Van Lente, Hekkert, Smits and Van Waveren (2003).
[221] Boon and Moors (2008); Smits (2002); Stewart and Hyysalo (2008); Verona et al. (2006).
[222] Burke, Rangaswamy and Gupta (2001); Nambisan (2002); Sawhney and Prandelli (2000).

policy[225] and science[226]. Thus, open innovation intermediaries help to overcome the innovators' reluctance towards contributing to an organizers innovation problem by acting as a trusted and unbiased third party in the relationship of innovators and organizers[227].

Table 2: Functions of open innovation intermediaries[228]

Dimension	Function	Description of activities
Connection	Gatekeeping and brokering	Linking organizers and innovators; facilitating knowledge flows; integrating knowledge from different domains
	Middle men between science policy and industry	Facilitating communication on a system level; Connecting on a system level
	Demand articulation	Connecting organizers and its customers
Collaboration and support	Knowledge processing and combination	Integrating knowledge; mobilizing research
	Commercialization	Supporting marketing, sales and funding
	Foresight and diagnosis	Aligning public research towards clients' needs
	Scanning and information processing	Scanning, scoping and filtering of external markets
Technological services	Intellectual property	Managing and controlling intellectual property
	Testing, validation and training	Testing, prototyping, diagnosing, analyzing and validating technology
	Assessment and evaluation	Assessing and evaluating technology
	Accreditation and standards	Consulting on standards and standard-setting
	Regulation and arbitration	Supporting formal, information and self-regulation as well as arbitration

Secondly, open innovation intermediaries provide *collaboration and support* services. They advise their clients on technological and managerial issues[229], foster in-house re-

---

[223] Benassi and Di Minin (2009); Chesbrough (2006b); Huston and Sakkab (2006); Klerkx and Leeuwis (2008), (2009); Winch and Courtney (2007).

[224] Becker and Gassmann (2006); Hansen, Chesbrough, Nohria and Sull (2000).

[225] Bessant and Rush (1995); Carlsson and Jacobsson (1997).

[226] Kodama (2008); Piore (2001); Stankiewicz (1995); Tether and Tajar (2008); Turpin, Garrett-Jones and Rankin (1996); Youtie and Shapira (2008).

[227] Hagel III and Rayport (1997); Sherry and Kozinets (2001).

[228] Building on Bessant and Rush (1995); Howells (2006); Lopez-Vega and Vanhaverbeke (2009).

search and the recombination of knowledge[230]. Furthermore, they provide marketing and sales support[231], help to identify market needs[232] and trends[233] and also screen markets for their clients [234]. Thirdly, open innovation intermediaries provide *technological services* by advising on intellectual property rights [235], standards [236], technology evaluation and staff training[237]. Table 2 summarizes the functions of open innovation intermediaries and outlines the typical activities performed to fulfill these functions. The following sections outline benefits and challenges that result from these functions.

2.3   Benefits of working with open innovation intermediaries

Compared to traditional innovation intermediaries, open innovation intermediaries work in a virtual environment, i.e. they use OIPs to integrate innovators into organizers' innovation processes. Working with open innovation intermediaries yields benefits for a client that apply to working with innovation intermediaries in general, as well as benefits that are specific to the virtual environment OIIs exploit. These benefits are introduced in the following.

In general, according to Bessant and Rush[238], open innovation intermediaries enable their clients to overcome a lack of organizational capabilities by contributing to their skills. The required organizational capabilities include the *recognition of requirements*, the *exploration, comparison and selection of options*, as well as the *acquisition, implementation and operation of a technology*. The contributions of OIIs to their clients' organizational capabilities are introduced in the following paragraph.

Open innovation intermediaries help their clients to *develop key management capabilities* to manage innovation projects and processes. Furthermore, they provide them with training on how to use innovation tools and methods. By *institution building* they mobilize a critical mass of knowledge that can be leveraged for the creation of innovations. By offering and transferring better management practices, they *reduce the risk of failure* while simultaneously *lowering cost*. Through direct

---

[229] Bessant and Rush (1995).
[230] Becker and Gassmann (2006).
[231] Bessant and Rush (1995); Howells (2006); Lichtenthaler and Ernst (2009).
[232] Bessant and Rush (1995).
[233] Van der Meulen and Rip (1998); Seaton and Cordey-Hayes (1993).
[234] Becker and Gassmann (2006).
[235] Benassi and Di Minin (2009).
[236] Howells (2006).
[237] Howells (2006).
[238] Bessant and Rush (1995).

outreach of open innovation intermediaries to third parties, *targeted support*[239] becomes possible resulting in higher efficiency. Furthermore, due to the *decentralized operation* when involving OIIs, less monitoring and control is required. These general contributions of open innovation intermediaries to their clients' organizational capabilities are summarized in Table 3.

Table 3: Contributions of OIIs to their clients' organizational capabilities[240]

Contribution	Description
Capability building	Building up management capabilities at client
Institution building	Mobilizing a critical mass of knowledge
Failure avoiding	Reducing the risk of failure through better management practices
Lowering cost	Lowering cost through better management practices
Targeting of support	Providing direct outreach to target audience
Decentralized operation	Decentralized management requiring less monitoring and control

Working in a virtual environment – as it is the case for open innovation intermediaries that use OIPs – results in additional benefits for a client. Four major areas are affected: the virtual environment enables new flexibility in *network access*, allows access to more and new *types of knowledge*, the OII can act as a *neutral third party*, and new *modes of knowledge absorption, integration, selection and implementation* are possible[241]. These areas are set out in the following paragraphs.

In the area of *network access*, the virtual environment allows to cost-effectively[242] create *direct ties*, i.e. direct relations, to potential innovators without facing the trade-off between media richness and physical reach that traditional innovation intermediaries have to deal with[243]. Thanks to a virtual environment, numerous innovators can be included in an innovation process[244]. They can participate and cooperate without regional limitations, exploiting an almost cost-free exchange of information[245]. Based on a self-selection mechanism of innovators[246],

[239] Burke et al. (2001).
[240] Building on Bessant and Rush (1995).
[241] See Verona et al. (2006) and Moeslein and Neyer (2009).
[242] Afuah (2003).
[243] Craincross (1997); Evans and Wurster (1999).
[244] Piller and Walcher (2006).
[245] Evans and Wurster (1999); Quinn (2000); Sproull and Kiesler (1991); Surowiecki (2005); Verona et al. (2006).

rather than screening[247] or pyramiding[248] to identify relevant innovators, an open innovation intermediary has the opportunity to access relevant innovators directly without the need for another intermediary like retailers in the client's sector or market research firms from the domain[249]. Furthermore, positive network effects can be realized[250]: Additional direct ties lead to the progressive decrease of costs to reach additional users[251], while additional users increase the value for existing users[252]. To sum it up, the number of direct ties to potential innovators can be increased. Apart from direct ties, *indirect ties* help to access more remote knowledge. A publicly accessible platform that accepts contributions from everybody [253] can encourage innovators to contribute, for instance, the partners a company's partner has on its parts[254]. Another method is to analyze existing knowledge in a partner's virtual community [255]. On the dissemination side, *structural autonomy*[256] allows the open innovation intermediary to sell information multiple times to different clients[257]. Structural autonomy is fostered by converging industries which results in new structural holes that can be leveraged by open innovation intermediaries[258].

The *type of knowledge* open innovation intermediaries generate can differ from traditional innovation intermediaries. The use of an open innovation intermediary helps to access *diverse knowledge* from different domains[259]. Companies hosting an OIP themselves tend to attract their own customers, which limits the accessible pool of knowledge[260]. This creates the risk of ignoring knowledge that does not fit current ideas of the organization[261] and, as a consequence, assuring oneself of one's own

[246] Piller and Walcher (2006).
[247] Morrison et al. (2004).
[248] Von Hippel et al. (2009).
[249] Nambisan (2002); Sawhney, Verona and Prandelli (2005).
[250] Downes and Mui (1998).
[251] Shapiro and Varian (1999).
[252] Gladwell (2000).
[253] Ruefli, Whinston and Wiggins (2001).
[254] Ahuja (2000); Gulati and Gargiulo (1999).
[255] Afuah (2003); Verona et al. (2006).
[256] Structural autonomy is a network property that implies that the intermediary has no structural holes (missing relationships between multiple parties in an industry) in the relationship at his own end while having rich structural holes at its clients ends. See Burt (1995).
[257] Werbach (2000).
[258] Linder, Jarvenpaa and Davenport (2003); Prahalad and Ramaswamy (2004); Verona et al. (2006).
[259] Quinn (2000); Sproull and Kiesler (1991); Surowiecki (2005); Verona et al. (2006).
[260] Nambisan (2002); Sawhney and Prandelli (2000).
[261] Christensen (2006).

mental models. This might result in a misinterpretation of market needs [262]. Contrarily, owing to their gatekeeping and brokering function, open innovation intermediaries can provide access to a magnitude of potential innovators from different domains, while, if needed, having the capability to address a specific customer segment[263]. This helps to overcome the perception bias due to selectively listening to one's own customers only [264]. Facilitating access to *social customer knowledge* is another capability of open innovation intermediaries. Customers interact online and influence each other when choosing a product[265]. This interaction can be analyzed using methods which might be hard to master for a non-specialized organization[266].

As open innovation intermediaries provide a *neutral third party* that represents the OIP organizer, i.e. their clients, customers are more willing to share knowledge (e.g. about their lifestyle and interests) with them than they would share directly with the organizer[267]. The reason for this is that customers tend to be less afraid of vested interest on a company's website or community than of a commercial exploitation of their knowledge. In another case, customers might not even be aware of a potential company that could address their need. They might thus rather address an untargeted, autonomous third party that forwards their need to an appropriate organization[268].

In the area of *knowledge absorption* open innovation intermediaries draw on multiple IT-based tools for open innovation[269]. In addition, they can use traditional methods like tracking analysis, surveys or conjoint analyses to create and absorb knowledge[270]. In contrast to traditional innovation intermediaries, open innovation intermediaries tend to generate and absorb new ideas and knowledge rather than focusing on the recombination of existing knowledge[271]. Through the capability of open innovation intermediaries to address a targeted audience, it becomes possible to

---

[262] Christensen and Bower (1996).
[263] Burke et al. (2001); Verona et al. (2006).
[264] Verona et al. (2006).
[265] Rheingold (2000); Rogers (2003).
[266] See for instance Verona and Prandelli (2002).
[267] Hagel III and Rayport (1997); Sherry and Kozinets (2001).
[268] Maes (1999).
[269] See section 1.3 and Arora, Fosfuri and Gambardella (2004); Lynn et al. (1996).
[270] Burke et al. (2001); Dahan and Hauser (2002).
[271] Hargadon and Sutton (1997); Hargadon (1998); Verona et al. (2006).

absorb specific knowledge, e.g. of a particular domain[272]. At the same time, knowledge absorption can be conducted rapidly, as virtual environments achieve a noticeable increase in the speed of interaction among the innovators, while they simultaneously document the innovation process and interim results[273].

Furthermore, in terms of *knowledge integration*, knowledge can be transmitted and shared more broadly[274]. While the formalized mechanisms on an OIP facilitate systematic information retrieval, awareness about available knowledge can be increased[275]. The informal character of, for instance, innovation communities permits a stronger social integration between the organizer and its customers, which allows for situated and distributed knowledge integration[276]. Both aspects facilitate the integration of the newly created knowledge[277].

In terms of *knowledge selection* open innovation intermediaries can leverage a broad user base to evaluate knowledge and knowledge combinations in order to identify promising instances that can be exploited. Covered by the term of open evaluation[278], open innovation intermediaries can draw on the wisdoms of the crowds to evaluate the vast amount of knowledge and select promising knowledge[279] without the perception bias as described above[280] or the need for many resources[281].

Finally, *knowledge implementation* is more difficult since particularly tacit knowledge is difficult to exchange in virtual environments[282]. This tacit knowledge is, however, important to promote an idea to an innovation. Open innovation intermediaries apply for instance innovation toolkits[283] and use prototypes[284] to transfer tacit knowledge from the innovators to the organizers. The previously stored knowledge can be recombined on demand for new purposes because it is easily retrievable[285]. As the knowledge codified on an OIP is transferrable and replicable in easier ways than traditionally codified knowledge, it can be offered to a broader set of

---

[272] Burke et al. (2001). See also the remarks on the type of knowledge created above.
[273] Moeslein and Neyer (2009); Sproull and Kiesler (1991).
[274] Wayland and Cole (1997).
[275] Verona et al. (2006).
[276] Brown and Duiguid (1991); Lave and Wenger (1991); Sproull and Kiesler (1991).
[277] Verona et al. (2006).
[278] Haller (2012); Moeslein et al. (2010).
[279] Datta (2007); Ebner, Leimeister, Bretschneider and Krcmar (2008); Surowiecki (2005).
[280] Burke et al. (2001); Nambisan (2002).
[281] Afuah and Tucci (2012); Gassmann, Sandmeier and Wecht (2006).
[282] Afuah (2003); Fahey and Prusak (1998); von Hippel and Katz (2002).
[283] Von Hippel and Katz (2002).
[284] Bullinger et al. (2011).
[285] Fahey and Prusak (1998); Ruefli et al. (2001).

potential clients. This increases the likelihood for OIIs to find one or more buyers[286]. Table 4 summarizes the major differences between open innovation intermediary-mediated, innovation intermediary-mediated and a company's direct access to external knowledge.

   To sum it up, from an innovation process perspective[287] open innovation intermediaries offer a different kind of strength than traditional innovation intermediaries. Open innovation intermediaries primarily support the *search* phase with a broad user based and access to diverse knowledge. Thus, they can create new knowledge and recombine existing knowledge on a broad basis. The same broad user base can be leveraged to *select* promising knowledge, which makes it possible to process a high amount of information with limited resources. Due to the virtual character of open innovation platforms and the codified insights, the knowledge is not directly transferrable in most cases, but also has to be translated to an organizer's context for implementation. Consequently, *implementation* of knowledge is harder for open innovation intermediaries than for traditional innovation intermediaries.

Table 4: Comparison of OII-mediated, innovation intermediary-mediated and direct access to external knowledge[288]

	Open innovation intermediary	Innovation intermediary	Direct access
Type of contact	Mediated	Mediated	Direct
Source of knowledge	Broad (from consumers and industries)	Limited (mainly from industries)	Limited to industry
Amount of processable knowledge	High	Medium to low	Low
Type of orientation	Network orientation	Client orientation	Firm orientation
Core competency	Search and select	Search and implement	Implement
Main limitation	Knowledge implementation	Network access	Network access

---

[286] Afuah (2003).

[287] An innovation process is split into the phases search, select and implement. For more details see section 1.2.

[288] Building on Terwiesch and Xu (2008); Verona et al. (2006).

According to the circumstances of open innovation intermediaries, their major strength lies in particular functions to support a client's innovation process[289]. Firstly, *gatekeeping and brokering* is facilitated by the broad source of knowledge open innovation intermediaries can draw on and the ability to process high amounts of knowledge. Secondly, consumer *demand articulation* is facilitated, as open innovation intermediaries can act as an independent trustworthy third party with access to customers and other potential innovators. Thirdly, *knowledge processing and combination* as well as *scanning and information processing* is facilitated by the broad access to diverse knowledge which is partly proactively submitted to the OII. This helps the open innovation intermediary to keep up with markets and it facilitates a distributed codification and integration of knowledge. Finally, *assessment and evaluation* can be carried out on a large and distributed scale.

In order to exploit these strengths, Verona et al. [290] propose that open innovation intermediaries should develop competencies in *tracking and profiling customers*, *creating incentive systems* and *managing two-way communication* to create social and individual knowledge. Also, they should have the ability to *analyze generated content* in order to recombine and transfer it to their client. Table 5 summarizes the skills that an OII should hold. Besides these benefits, working with open innovation intermediaries holds challenges as described in the following.

Table 5: Competencies that open innovation intermediaries should develop

Competency
Tracking and profiling customers
Creating incentive systems
Managing two-way communication
Analyzing generated content

2.4   Challenges of working with open innovation intermediaries

Besides new opportunities, particular problems arise when working with open innovation intermediaries. Firstly, *selecting the right problems* to post to an OIP is a challenge[291]. Problems that are to be solved with an open innovation intermediary

---

[289] See section 1.2.
[290] Verona et al. (2006).
[291] Sieg et al. (2010); Terwiesch and Xu (2008).

should share two characteristics. They have to be solvable by externals and they have to be revealable to the world. Sieg et al. [292] find that their case companies tend to post complex, commercially highly valuable problems to OIPs they cannot solve themselves. They hope to find professionals who can solve their problem with knowledge from a different domain. The benefit of these 'holy grail' problems is that very little is known about them so far. Thus, little internal knowledge has to be revealed while these problems yield the chance to create high revenues. But also difficulties with lower complexity are passed on to address tinkerers and hobbyists for instance[293]. In any case, results of the external problem solving process need to be reintegrated into the internal problem solving processes, which has to be considered right from the outset. Revealing problems to the world has two downsides. On the one hand, problems need to be explicitly stated so they can be understood by outsiders. This yields the risk of trade secrets being revealed or patenting becoming an issue. Furthermore, sensitive information that constitutes a competitive advantage might have to be revealed in order to find a solution[294]. Terwiesch and Xu[295] provide a framework, which is not further elaborated here, to decide based on the type of the innovation project whether the innovation project should be conducted internally, in cooperation with an open innovation intermediary, or by the company itself using open innovation. In any case, the selection of appropriate problems requires some internal work.

Once the right problems are selected, *formulating problems* appropriately is the next challenge. The sole information for innovators on an OIP is the formulated problem statement. This statement has to enable innovators to *recognize similarities* between the problem and a solution from their domain. This is challenging, as in regular work practice problems are formulated circularly: Once a part of a problem is solved, the formulation of the problem is refined and so on. When using OIPs, the problem formulation is linear: problems are posted to the OIP only once. A refinement generally does not take place on a frequent basis[296]. The second challenge in formulating problems is the overcoming of the *specific language* a problem poster might apply. Consider, for instance, an employee that interacts with his colleagues. They have a common understanding of the vocabulary they use, which is not familiar

---

[292] Sieg et al. (2010).
[293] Hallerstede et al. (2010).
[294] Henkel (2006).
[295] Terwiesch and Xu (2008).
[296] Sieg et al. (2010).

to externals[297]. Thirdly, as employees might already have parts of a solution to a problem, they might tend to formulate the problem according to the missing pieces of their own solution rather than according to the *ultimate target*, which could also be reached with a completely different solution[298].

Third, *employees' reluctance* at the client's organization due to new work practices has to be overcome. This applies to two areas: The *posting of problems* on OIPs and the *internalization of externally generated knowledge*. In a typical work practice, researchers within a company discuss a problem with colleagues who know the problem's context and share a common language. Due to this fact, explicit problem statements, which are required to post a problem to an OIP, are rare. In addition, researchers might already have solutions to parts of a problem. As a consequence, they tend to ask only for the missing parts towards a possible solution and not the entire problem, which hampers radically different approaches to solve the problem. These two factors make it hard, time-consuming and tedious to post problems to an OIP, as these challenges have to be overcome to create a precise problem description externals can work with[299]. At the other end of the innovation process, due to the not-invented-here syndrome[300], implementation of ideas generated in collaboration with open innovation intermediaries might be hindered. To overcome employees' reluctance, clients should integrate and train them to feel comfortable with these new work practices[301].

Finally, as open innovation intermediaries rely on software, i.e. OIPs, to integrate innovators, they face limitations due to the use of *software-mediated knowledge transfer*[302]. The software predefines certain options to interact with it, making it thus impossible to react to situational factors. An OIP might, for instance, only offer to submit contributions in writing, while the contributor might be willing to share richer information like drawings, pictures or prototypes. But even if multiple options to submit contributions are offered, complex contributions are not easily transferrable using software-mediated communication. Two consequences arise. Firstly, open innovation intermediaries should implement a sound technological basis that

---

[297] Bjoerkdahl and Linder (2010); Sieg et al. (2010).
[298] Sieg et al. (2010).
[299] Sieg et al. (2010).
[300] Biemans (1991); Katz and Allen (1982); Piller and Walcher (2006).
[301] Pittaway et al. (2004).
[302] Daft, Lengel and Trevino (1987).

complements the objectives[303]. Secondly, the technology alone does not suffice but has to be complemented by human agents, who support knowledge transfer[304].

Having introduced open innovation platforms (*chapter 1*) and open innovation intermediaries who use OIPs (*chapter 2*), the third facet in OIP lifecycle management that links the two first facets, namely the process of OIP lifecycle management, is addressed in the following *chapter 3*.

---

[303] Pittaway et al. (2004).
[304] Datta (2007); Jahn (2005); Lichtenthaler and Ernst (2008).

3   A lifecycle model for open innovation platforms

OIPs have to be designed and managed[305]. In order to enable an OIP's success, an integrated view on the design and management is crucial[306]. An OIP lifecycle manager, like for instance an open innovation intermediary, can draw on software lifecycle models in order to structure its work[307]. This chapter aims at selecting an appropriate lifecycle model for OIP lifecycle management and at adapting it to the field. These foundations define a model of OIP lifecycle management that is used to structure the case analysis in Part III. In addition to that, a model of OIP lifecycle management allows providing guidelines for research and practice. These objectives are to be tackled as follows.

Firstly, the contributing literature (*section 3.1*) and the need for a lifecycle approach for OIPs is outlined (*section 3.2*). Additionally, different types of software development projects are highlighted to derive implications for an appropriate software lifecycle model for OIP projects (*section 3.3*). A comparison of popular software lifecycle models follows. The aim of this comparison is to assess their suitability to structure the design and management of OIPs (*section 3.4*). Based on the comparison, a model is selected (*section 3.5*) and adapted to the field, resulting in the OIP lifecycle model (*section 3.6*).

3.1   Relevant research streams for this chapter

Two major streams in the information systems literature cover information systems (IS) design[308] and information systems management[309]. While the IS design stream deals with tools[310], techniques[311] and processes[312] to create new software, the IS management streams investigates how to coordinate this work with lifecycle[313] and

---

[305] OGC (2002). Dalcher (2002) for instance uses the terminology of design and management for software lifecycles as well.
[306] See section 3.2 below.
[307] See Part I.1.
[308] Abrahamsson, Salo, Ronkainen and Warsta (2002); Bell (2009); Sommerville (2011).
[309] Bon, Pondman and Kemmerling (2002); Taylor (2003).
[310] Kernighan and Plauger (1976); Loeliger and McCullough (2012).
[311] Cherns (1976); Mayhew (1991).
[312] Hickey and Davis (2002); Kotonya and Sommerville (1998); Lowe and Henderson-Sellers (2001).
[313] Benediktsson, Dalcher and Thorbergsson (2006); Hoffmann and Beaumont (1997); Shaw (2007).

role models[314], how to manage software[315] and how to design a purposeful software landscape[316].

This chapter takes a software design and management perspective[317]. The two streams of software design and software management are integrated using a software lifecycle model[318]. In order to link the IS view with OIPs, the impact of findings from open innovation literature on the software lifecycle model is explained. Thus, this chapter also refers to open innovation literature as described in section 1.1. The following elaborates why a lifecycle approach is needed for OIP design and management.

3.2   Need for a lifecycle approach

Current research[319] and practice[320] stress the importance of pre-launch and post-launch phases of software projects. *Pre-launch phases* include all activities that are processed prior to an OIP's launch. It hence includes requirements engineering as well as designing and implementing the OIP and its architecture. Thus, the social and technical subsystem of an OIP[321] is defined in the pre-launch phases. *Post-launch phases* include all activities that are processed after an OIP is built, namely the OIP rollout, maintenance and changes to it. The focus lies on the establishment of the intended social subsystem and the identification of improvements for an OIP. The combination of the pre- and post-launch phases of an OIP, namely

> *an OIP lifecycle, is defined as all activities that are required to design (pre-launch phases) and manage (post-launch phases) an open innovation platform.*

It thus covers the "period of time that begins when a software product is conceived and ends when the software is no longer available for use"[322]. This definition does not imply any process, but is a general definition of an OIP lifecycle that includes for

---

[314] Thayer (2000).

[315] Bon et al. (2002); OGC (2002); Taylor (2003).

[316] Addicks and Steffens (2008).

[317] Whereas software refers to „computer programs, procedures, and possibly associated documentation and data pertaining to the operation of a computer system" (IEEE 1990, p. 66) an application refers to the parts of a software that a user can execute. Although both terms are used in their context, they are considered being synonyms in the broader sense of software.

[318] Guntamukkala, Wen and Tarn (2006); OGC (2002).

[319] For instance Pham (1999); Sommerville (2011).

[320] For instance Bell (2009); Orlikowski et al. (1995).

[321] See section 1.4.

[322] IEEE (1990), p. 68. In this thesis, the terms software development project and software development are used synonymously to the lifecycle of an OIP.

instance incremental improvements to an OIP. An OIP's lifecycle will later on be structured based on a software lifecycle model[323].

The importance of the *pre-launch phases* can be read from a study by KPMG[324]: 62% of the investigated knowledge management projects, which have similar project characteristics like OIP projects, failed right after the initial launch, which indicates flaws in the pre-launch phases. Particularly for OIPs, the pre-launch phases are crucial, since they, for instance, include the design of the social subsystem. Consequently, many design-based decisions have to be taken[325].

Nonetheless, the *post-launch phases* are important as well. Current paradigms of managing web 2.0 software include the perpetual-beta-paradigm[326]. This paradigm requires a constant improvement and many small releases of a software. Particularly in the case of OIPs, post-launch phases require active management and improvement of the application, for instance, to motivate users to participate[327].

So far, little is known about an integrated approach to design and manage OIPs[328]. As an OIP is software[329], this thesis draws on *prescriptive software lifecycle models*[330] to identify an appropriate software lifecycle management model for OIPs[331].

Prescriptive software lifecycle models provide "guidelines or frameworks to organize and structure how software development activities should be performed, and in which order"[332]. Prescriptive software lifecycle models can thus provide hints on the selection of activities in an OIP lifecycle, which is required to derive guidelines on OIP lifecycle management. In contrast to software *lifecycle* models, software *development process* models "represent a networked sequence of activities, object, transformations, and events"[333]. Therefore, they cover more details than a software lifecycle model. As this additional degree of detail is not required for the intended

---

[323] See section 3.6.
[324] KPMG (2001).
[325] Bullinger and Moeslein (2011); Hallerstede and Bullinger (2010).
[326] Hoyer, Schroth, Stanoevska-Slabeva and Janner (2007).
[327] Adamczyk, Boehler, Bullinger and Moeslein (2011); Fueller et al. (2008); Koh, Kim, Butler and Bock (2007).
[328] Iriberri and Leroy (2009).
[329] See section 1.2.
[330] Prescriptive software lifecycle models prescribe how new software should be developed, whereas descriptive models describe how a particular software was actually developed (Scacchi 2001).
[331] Elliott (2004). Iriberri and Leroy (2009) for instance follow a similar approach in adapting an IS model to online communities.
[332] Scacchi (2001), p. 3.
[333] Scacchi (2001), p. 4.

analysis, both terms are subsumed under the term software lifecycle model. In general, software lifecycle models serve to plan, organize, staff, coordinate, budget and direct software projects[334]. There are different types of software development projects. Since the selection of a suitable software lifecycle model depends on the type of software development project, the different software development project types are set out in the following.

3.3   Types of software development projects

Boehm[335] distinguishes three types of software development projects based on the criteria *team size, experience of the team members, rigidity of requirements* and *target of the software development*:

Embedded mode projects require a comprehensive assessment of rigid require-ments in the beginning of a software lifecycle. This is due to a "strongly coupled complex of hardware, software, regulations and operational procedures" [336] a software operates in (or: 'is embedded in'). This allows for large development teams with experienced and un-experienced developers who parallelize development. Requirements cannot be negotiated (or only slightly), as changes to other systems are expensive or not even possible. The aim is to build a stable and reliable software. For this type of project, heavyweight software lifecycle models are appropriate.

Organic mode projects have opposing project characteristics. Relatively small project teams work in a familiar IT environment. Developers are highly skilled and know the topics. Requirements are neither fixed nor known completely from the beginning and can be negotiated over time. The focus lies on functionality that has to be developed during the project. For this type of project, lightweight software lifecycle models are appropriate.

Semidetached mode projects represent an intermediate stage between the two preceding project types, representing a mixture of the organic mode and embedded mode characteristics. Some requirements are fixed and some are not. Mid-size teams comprise experienced and un-experienced developers who might be experts for parts of the system. For this type of project, lightweight and heavyweight software lifecycle

---

[334] Scacchi (2001).
[335] Boehm (1981). For an updated typology that takes current developments in web 2.0 (O'Reilly 2005) into consideration see Boehm and Turner (2003). However, the basic propositions of the three software development project types, as introduced at this point, are still valid.
[336] Boehm (1981), p. 79.

models can be applied. The decision depends on the details of the project's characteristics[337].

This distinction of project types is relevant when looking into the *coordination* of work of an OIP lifecycle manager, like for instance an open innovation intermediary. Following this typology, OIP projects tend to reflect organic mode projects with volatile requirements and small and knowledgeable teams [338]. Consequently, open innovation intermediaries should apply a predominantly agile software development approach to coordinate their work.

This typology does actually not affect the suitability of a lifecycle model to *structure* the lifecycle of an OIP. The software lifecycle model shall primarily provide guidelines on which activities to perform and to focus on in an OIP's lifecycle. The following section derives criteria for such a suitable software lifecycle model to *structure* an OIP's lifecycle and selects a software lifecycle model based on these criteria. An OIP lifecycle manager can still apply a different model to coordinate its actual day to day work than the model to structure the OIP lifecycle.

3.4   Comparison of software lifecycle models

Three criteria are applied to select an appropriate software lifecycle model to structure the design and management of OIPs. Firstly, the selected software lifecycle model should have an integrated and *balanced view* on pre- and post-launch phases of an OIP's lifecycle[339]. Using a balanced software lifecycle model ensures that all phases of an OIP's lifecycle are equally considered and no activities are omitted or overemphasized in the analyses to come. As current knowledge does not provide any hint on the comparative importance of pre-launch and post-launch phases for OIPs, they have to be treated and considered as equally important. Secondly, the selected lifecycle model should provide an appropriate *level of abstraction*. Its phases should neither be too general for them to provide some structuring of an OIP lifecycle, nor should they be too detailed in order to be adaptable to different OIP projects. In addition, a too detailed structure would be unhandy for practitioners, since too many phases would have to be considered. Thirdly, all relevant activities in an OIP's

---

337 Boehm (1981). For an updated typology that takes current developments in web 2.0 (O'Reilly 2005) into consideration see Boehm and Turner (2003). However, the basic propositions of the three software development project types, as introduced at this point, are still valid.

338 See Diener and Piller (2010).

339 For details on the pre- and post-launch phases and reasons that require a balanced view see section 3.2.

lifecycle should be covered by the software lifecycle model in order to ensure *comprehensiveness* of the OIP lifecycle model.

The following introduces common[340] software lifecycle models that might potentially serve to structure an OIP's lifecycle. The software lifecycle models are evaluated in a two-step process applying the criteria mentioned before[341]. In *step one*, their general suitability with regards a *balanced view* on pre- and post-launch phases and an appropriate *level of abstraction* is verified. The qualifying models are tested in *step two* concerning the third criterion of *comprehensiveness*. Applying this two-step process allows running the intricate test with regard to comprehensiveness only for qualifying models that are potentially suitable for structuring an OIP's lifecycle.

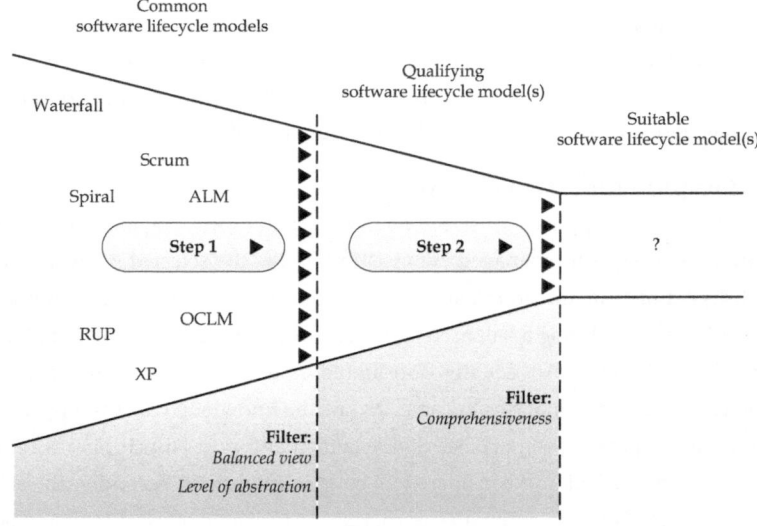

Figure 10: Two-step process to select an appropriate lifecycle model for OIPs

In order to analyze a variety of different software lifecycle models, representatives of *agile, plan-driven* and *incremental models* are selected for the evaluation. *Plan-driven models* plan and specify a software in advance and implement it based on this plan. They are particularly suited for software with fixed requirements and high stability and safety requirements. *Agile models* contrarily plan incrementally and adapt to

---

[340] Benediktsson et al. (2006); Guntamukkala et al. (2006).
[341] See Figure 10.

changing requirements[342]. In practice, a manager of a software development project will need to find a balance between agile and plan-driven models[343]. This gap between plan-driven and agile models is addressed by *incremental software development models*[344].

According to Boehm and Turner's[345] criteria, OIP development should be *coordinated* using a predominantly agile approach, as an OIP project resembles an organic mode project: not all requirements are known from the very beginning and there are only few systems that might have to connect to an OIP. However, *structuring* an OIP's lifecycle might draw on all models[346].

The following subsections introduce typical[347] plan-driven (*waterfall model*), incremental (*spiral model, rational unified process, application lifecycle management, online community lifecycle model*) and agile (*extreme programming, Scrum*) software lifecycle models[348] and evaluates their suitability to structure an OIP's lifecycle.

3.4.1 Waterfall model

The waterfall model was originally developed by Royce[349] and has been adapted and improved ever since[350]. Its original notion stems from the finding according to which the two-step process of analysis and coding - commonly applied during that time - suffices for small internal software development projects, but not for larger software development projects. Royce added two preceding phases to identify system and software requirements, one intermediary phase to define the program design (architecture) and two succeeding phases to conduct testing to the two-step process. He also included operations that define a more comprehensive process for software development to allow a better utilization of programming resources[351]. This model relies on documentation and coordination. Iterations of phases are allowed to adapt to changes.

---

[342] Boehm and Turner (2003).
[343] Sommerville (2011).
[344] Cockburn (2008); Larman and Basili (2003). Iterative models, as another type of agile software development models, are not distinguished in this thesis and – as commonly done (Larman and Basili 2003) - subsumed under incremental software development models.
[345] Boehm and Turner (2003).
[346] See section 3.3.
[347] Benediktsson et al. (2006); Guntamukkala et al. (2006).
[348] For a classification and comparison of these and further software lifecycle models see Benediktsson et al. (2006); Guntamukkala et al. (2006); Sasankar and Chavan (2011).
[349] Royce (1970).
[350] Scacchi (2001).
[351] Royce (1970). See Figure 11.

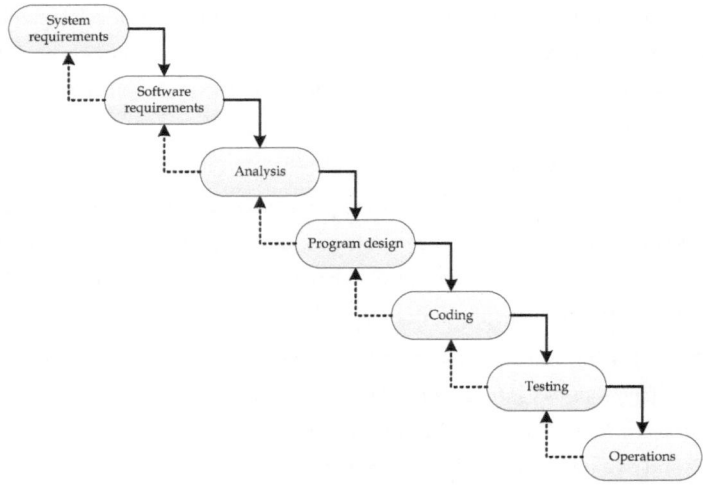

Figure 11: The waterfall model[352]

In this model, the phases system requirements to testing are part of the pre-launch phases[353]. Only one phase, namely operations, takes care of the post-launch activities. Thus, the waterfall model has a strong emphasis on the pre-launch phases of a software lifecycle and does not fulfill the requirement of a *balanced view*. The same objection arises for the V-Model[354], which is an evolution of the waterfall model[355]. It is thus not further considered at this point. Although the spiral model also represents an advancement of the waterfall model, it is analyzed in the following subsection, since it is a more lifecycle oriented approach than the waterfall model and V-Model[356].

3.4.2   Spiral model

The spiral model is one of the pioneers of incremental software lifecycle models[357]. It is a risk-driven approach that integrates the benefits of multiple software development models known at that time. Its four quadrants represent the activities during the software development process, i.e. determining objectives, evaluating

---

[352] Royce (1970).
[353] For a definition of pre-launch and post-launch phases see section 3.2.
[354] Broy and Rausch (2005).
[355] Boehm (1979); Broy and Rausch (2005); Droeschel and Wiemers (1999); Hoehn and Hoeppner (2008).
[356] Boehm (1988).
[357] Larman and Basili (2003).

alternatives, development and planning for the following phase[358]. These are processed iteratively, with a strong emphasis on risk analysis and coping with risk[359].

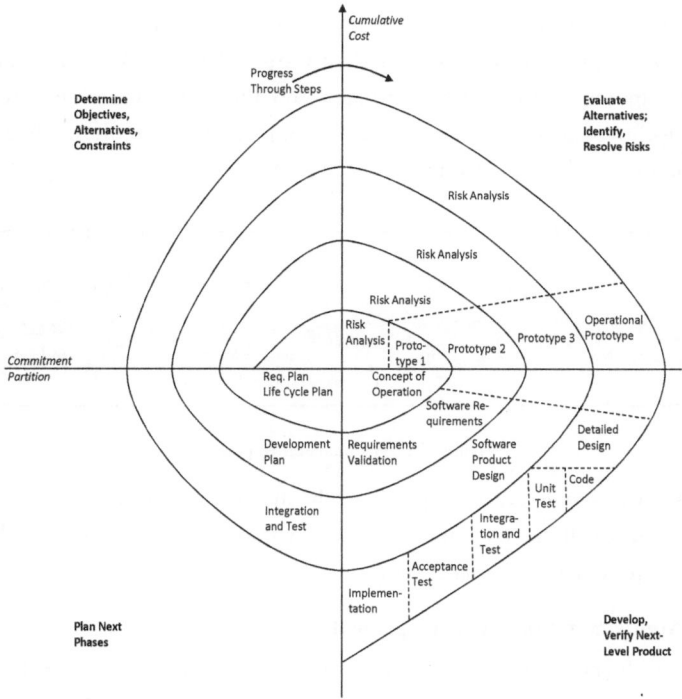

Figure 12: The spiral model[360]

Although this model works with frequent prototypes, it emphasizes the pre-launch phases and does not consider the post-launch phases. Therefore, it is not suitable for structuring an OIP's lifecycle, as it does not fulfill the requirement of a *balanced view*.

3.4.3 Rational unified process

The rational unified process (RUP) is derived from the unified modeling language (UML) and the unified software development process[361]. RUP distinguishes between a *dynamic* (phases of the model), a *static* (process activities) and a *practice* (good

---

[358] See Figure 12.

[359] Boehm (1988); Boehm et al. (1998).

[360] Boehm (1988).

[361] Arlow and Neustadt (2005); Rumbaugh, Jacobson and Booch (1999).

practices for the processes) *perspective*. The dynamic perspective might serve as an approach to structure an OIP's lifecycle, as it provides a structuring of a software's lifecycle phases. The dynamic perspective's phases *inception, elaboration, construction* and *transition* may be enacted in an iterative way as well as in an incremental way[362]. During *inception*, the business case for a software is defined. In the *elaboration* phase an understanding of the problem domain and the software architecture is built. *Construction* involves system design, programming and testing. In the final *transition* phase, the software is deployed to its users[363].

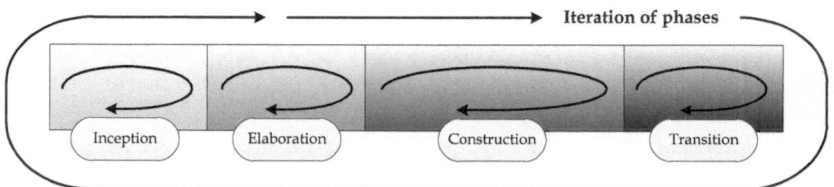

Figure 13: The rational unified process[364]

Although this model considers post-launch phases like the transition to the user environment, it still has its emphasis on the pre-launch phases and does hence not fulfill the requirement of a *balanced view*.

3.4.4   Application lifecycle management

Application lifecycle management (ALM) was created by the British Office of Government Commerce (OGC) with the aim to build a balanced lifecycle approach for software development projects in cooperation with the British government[365]. Within ALM, *application creation* and *service management* are distinguished. Whereas *application creation* refers to the ALM phases *requirements, design* and *build, service management* refers to the phases *deploy, operate* and *optimize* in a software's lifecycle[366]. The lifecycle of an application begins with the gathering of functional and non-functional *requirements*[367]. In the *design* phase, these requirements are translated into feature specifications. In the *build* phase the software and its architecture are realized.

---

[362] See Figure 13.
[363] Kruchten (2003); Sommerville (2011).
[364] Sommerville (2011).
[365] OGC (2002; 2011).
[366] See Figure 14.
[367] For a definition of functional and non-functional requirements see Bass, Clements and Kazman (2003); Behkamal, Kahani and Akbari (2009); Mich, Franch and Gaio (2003).

New components are purchased or built and subsequently integrated and tested. Once the system is built, the *deploy* phase starts. Therefore, the (changed) architecture has to be implemented on existing systems and the software has to be made available. During *operation*, support has to be given to the users and changes in the requirements have to be captured. The final phase in the ALM circle is the *optimize* phase. During this phase the results from operations are analyzed. Therefore, feedback has to be collected from the users and other means of evaluation have to be taken. The phases do not necessarily run consecutively, but can overlap due to parallel circles (e.g. multiple changes are implemented at the same time) or iterations (e.g. another circle starts before the prior one finished or two or more process steps have to be iterated).

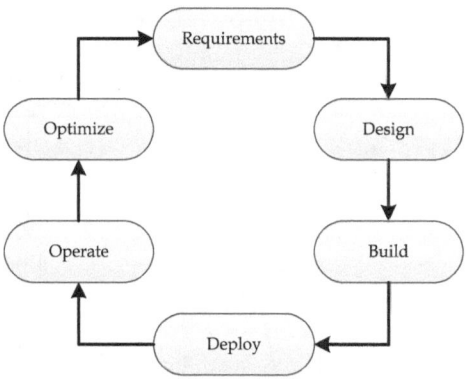

Figure 14: The application lifecycle management model[368]

ALM offers a *balanced view* on the pre-launch (requirements, design, build) and post-launch (deploy, operate and optimize) phases. It additionally represents an appropriate *level of abstraction* that allows for a structured analysis of an OIP's lifecycle while not getting lost in too detailed phases. Thus, ALM is potentially suitable for structuring an OIP's lifecycle and considered a qualifying software lifecycle model.

---

[368] OGC (2002).

3.4.5   Online community lifecycle model

Iriberri and Leroy [369] developed a model to describe the lifecycle of online communities based on a generic software lifecycle model. They distinguish the phases *inception, creation, growth, maturity* and *death*[370]. During *inception*, the idea for an online community evolves. In the *creation* phase, the technology is gradually incorporated and users join the community. During *growth* the culture and identity of the community are established and roles are distributed idiosyncratically, while formal structures and reward systems are implemented as the community reaches *maturity*. The community lifecycle iterates until a community's *death* due to a lack of participation or quality contributions.

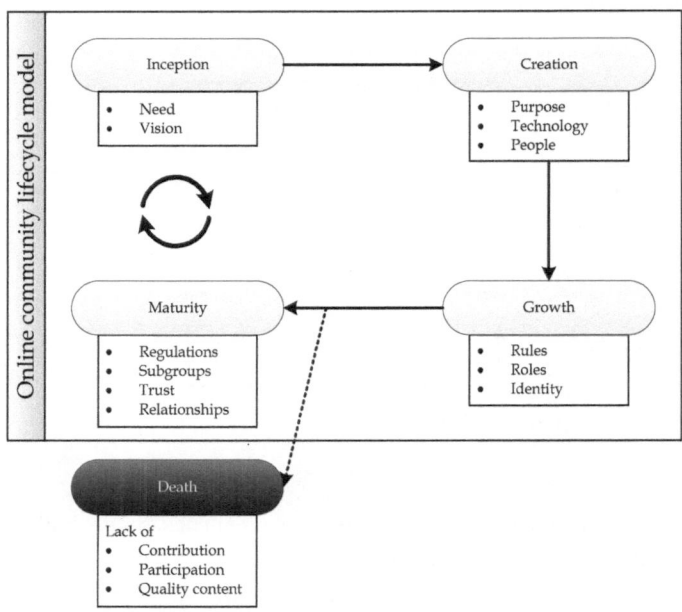

Figure 15: The online community lifecycle model[371]

This model has a strong emphasis on post-launch phases and the social subsystem of an OIP while simultaneously neglecting the design of the technical subsystem. Thus,

---

[369] Iriberri and Leroy (2009). Although the model is not an original IS model as claimed in section 3.2, it is included due to its potential fit for an OIP lifecycle and its at least partial origins in IS.

[370] See Figure 15.

[371] Iriberri and Leroy (2009).

it is not suited to structure an OIP's lifecycle, as it does not fulfill the requirement of a *balanced view*.

3.4.6   Extreme programming

Extreme programming (XP) is an agile method developed by Beck[372] that builds on frequent, early and continuous iterations which each process waterfall model-like phases[373]. The phases are *analysis, design, implement* and *test*. The incremental planning builds up towards a global plan during the runtime of a project. XP comprises a set of values, principles, activities and techniques to coordinate work and define procedures. Clients are continuously integrated into the process of software development[374].

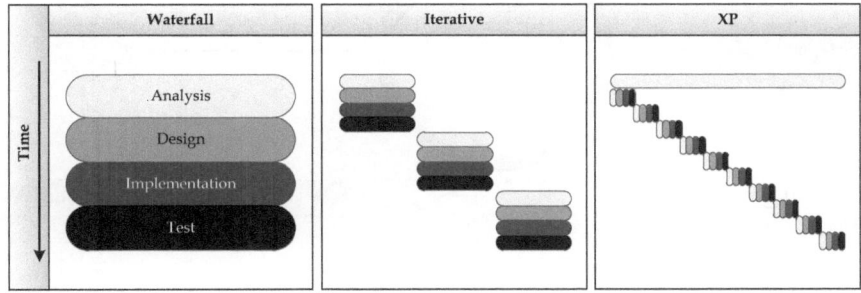

Figure 16: Extreme programming[375]

Due to its origin in the waterfall model, XP shares the same shortcomings concerning applicability to an OIP's lifecycle. Like the waterfall model, XP has a strong emphasis on the pre-launch phases and does not fulfill the requirement of a *balanced view*. It is thus not suited to structure the lifecycle of OIPs.

3.4.7   Scrum

Scrum is an agile software development model by Beedle et al.[376] who developed it based on their project experience with Takeuchi and Nonaka[377]. Scrum is an approach that rather manages complexity than a lifecycle model. Scrum consists of *roles*, a *flow*

---

[372] Beck (1999).
[373] For details concerning the waterfall model see subsection 3.4.1.
[374] Beck and Andres (2004).
[375] Beck (1999).
[376] Beedle et al. (1999).
[377] Takeuchi and Nonaka (1986).

and *artifacts*. The *Scrum-flow* combines the roles and artifacts in a sequence[378]. The software's vision is translated into a product backlog that contains requirements. The product backlog is split up into sprint backlogs which summarize the tasks for an upcoming sprint. The backlogs are frequently revisited. The result of each sprint is a finished functionality that can be delivered to the client[379].

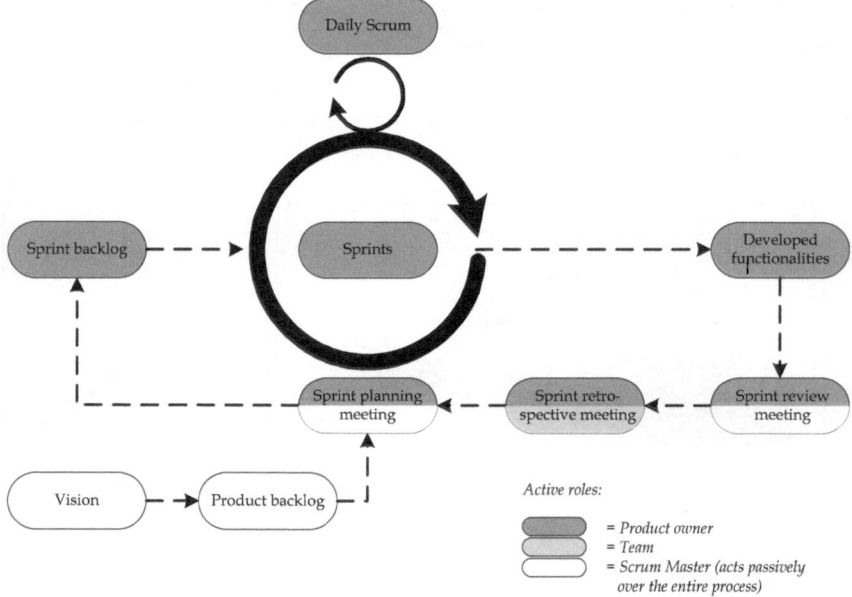

Figure 17: The Scrum-flow[380]

Scrum is a software development model that focuses on the pre-launch phases of software. The steps of the Scrum-flow are on different layers of abstraction. The post-launch phases are almost not represented at all. Therefore, it is not suitable for analyzing an OIP's lifecycle, as it does not meet the requirement of a *balanced view*. The following section summarizes the findings for the introduced software lifecycle models in order to select a suitable one (step 1) and test it with regard to its *comprehensiveness* in the second step (step 2).

---

[378] See Figure 17.
[379] Schwaber (2004).
[380] Schwaber (2004).

3.5   Selection of a software lifecycle model to structure an OIP's lifecycle

Table 6 summarizes the evaluation of software lifecycle models as presented in section 3.4. ALM is the only software lifecycle model that fulfills the defined requirements concerning a *balanced view* on pre-launch and post-launch phases and an appropriate *level of abstraction* for a software lifecycle model to structure an OIP's lifecycle. Thus, its suitability is further verified concerning the third criterion, *comprehensiveness*, of the software lifecycle model.

Table 6: Comparison of software lifecycle models to structure an OIP's lifecycle

Software lifecycle model	Balanced view on pre- and post-launch phases	Appropriate level of abstraction	Comprehen- siveness	Suitability
Waterfall model	Strong emphasis on pre-launch phases	Appropriate	Not evaluated	Not suitable
Spiral model	Not considering post-launch phases	Too detailed	Not evaluated	Not suitable
Rational unified process	Emphasis on pre-launch phases	Appropriate	Not evaluated	Not suitable
Application life-cycle management	Balanced pre- and post-launch phases	Appropriate	Given	Suitable
Online community lifecycle model	Emphasis on post-launch phases; neglecting technical aspects	Appropriate	Not evaluated	Not suitable
Extreme programming	Strong emphasis on pre-launch phases	Appropriate	Not evaluated	Not suitable
Scrum	Almost not considering post-launch phases	Varying; not appropriate	Not evaluated	Not suitable

The comprehensiveness of ALM has to be verified in order to ensure that all potentially relevant activities in an OIP's lifecycle are covered by it. Scacchi[381] identifies twelve typical activities that are part of software lifecycle models in his review on the same. Table 7 maps those activities to the six ALM phases in order to ensure that all activities are covered in ALM.

All activities identified by Scacchi[382] are mapped to an ALM phase, while simultaneously all ALM phases represent at least one activity. Thus, ALM covers all

---

[381] Scacchi (2001).
[382] Scacchi (2001).

relevant activities in a software lifecycle while not putting any focus on unnecessary details. To sum it up, ALM proves to be suitable for structuring the design and management of OIPs[383]. For this reason, this thesis builds on ALM to structure an OIP's lifecycle.

Table 7: Mapping of typical software lifecycle activities to the six ALM phases[384]

ALM phase	Software lifecycle activity	Description
Requirements	System initiation and planning	Plan what the system replaces in or supplements to current processes
	Requirements analysis and specification	Gather and manage requirements for the new software
Design	Functional specification or prototyping	Identification and formalization of attributes, relationships and other components
	Partition and selection	Subdivide software into manageable pieces and build vs. buy vs. reuse decision
	Architectural design and configuration specification	Define interfaces between subsystems, components and modules
	Detailed component design specification	Define procedural methods
Build	Component implementation and debugging	Convert specifications into source code and validate operation of subsystems
	Software integration and testing	Validate overall integrity of the software and verify accordance with requirements
Deploy	Documentation revision and system delivery	Create structured user guides and software documentation
	Deployment and installation	Provide guidelines for integrating the software to existing IT landscapes and configure corresponding systems
Operate	Training and use	Support and train users to use the software appropriately
Optimize	Software maintenance	Sustain operation of a software and implement repairs and enhancements

---

[383] See Table 6.
[384] Based on OGC (2002); Scacchi (2001).

In the following, ALM is introduced in the light of the social as well as the technical perspective of OIPs, as discussed in open innovation literature. This way, activities that are particular to OIPs are compared to the traditional notion of ALM.

3.6 The lifecycle of open innovation platforms according to ALM

ALM is part of the ITIL (information technology infrastructure library) framework, which was created by the British Office of Government Commerce in the 1980s[385]. ITIL is a collection of best practices to provide cost-effective IT services from an IT provider to its customer while granting an adequate quality[386]. ITIL's focus lies on the alignment of IT services and business needs. Today, ITIL has become a de-facto standard in IT management[387]. ITIL and, as a part of it, ALM contain general guidelines that are not specific to a field or organization. Although some authors stress that there is no need to treat the lifecycle of different types of applications individually[388], I follow the dominant line of argumentation by authors who ask for a treatment based on the application type[389]. Therefore, in the following, ALM is adapted to the field of OIPs while comparing it to the original notion of ALM. The adaption is structured along the six phases of ALM: *Requirements, design, build, deploy, operate* and *optimize*[390]. The result of this adaption is the OIP lifecycle model, as summarized in chapter 4.

3.6.1 Requirements phase

During the requirements phase, the requirements for an application are engineered. According to the IEEE standard software engineering terminology, a requirement is "(1) a condition or capability needed by a user to solve a problem or achieve an objective; (2) a condition or capability that must be met or possessed by a system or system component to satisfy a contract, standard, specification, or other formally imposed documents; (3) a documented representation or a condition or capability as in (1) or (2)"[391]. Requirements should focus on business needs[392]. Requirements engineering takes care of eliciting, specifying, verifying, validating, modeling,

---

[385] OGC (2002).
[386] Oecking and Degenhardt (2011).
[387] Hochstein, Zarnekow and Brenner (2004); Oecking and Degenhardt (2011).
[388] For instance Holck (2003).
[389] For instance Ahmad et al. (2005); Escalona and Koch (2003); Fraternali (1999); Pressman (1998).
[390] For a summary of the phases, see the above introduction to ALM in subsection 3.4.4.
[391] IEEE (1990), p. 62.
[392] OGC (2002).

analyzing, discussing and managing requirements in a structured process[393]. In traditional software development, requirements engineering is a core task to facilitate a software's success[394].

According to the OGC [395], six types of requirements are distinguished. *Functional requirements* request support for a specific business function. *Manageability requirements* address service management requirements and the need for a responsive, available and secure application that can be deployed and maintained. *Usability requirements* ensure easy usability for end users. *Architectural requirements* define required changes for or an adaption to existing architectures. *Interface requirements* assess connections to other existing applications and finally *service level requirements* define the intended performance and quality of the application. All but the functional requirements are subsumed under *non-functional requirements.*

While requirements can be gathered using different methods, popular requirements elicitation methods include scenario analyses, focus groups, modeling and joint application design[396]. Methods can complement or substitute each other[397]. The selection of an appropriate method[398] depends on the context of a project.

As OIPs are web-based software, characteristics for the requirements phase of this type of software apply. First of all, usability and user experience are key success factors for web-based software[399]. Secondly, as OIP projects are organic mode projects, functional requirements are of particular interest[400]. In the case of OIPs, requirements are potentially elicited continuously, as they evolve and as users are not necessarily known in advance[401]. The gathering of requirements is based on the purpose of the OIP. An OIP's purpose defines the target audience and the intentions of the organizer which have a direct impact on the OIP's design[402]. Requirements are to be gathered for both, the social and the technical subsystem of an OIP.

---

[393] Ebert (2008); Kotonya and Sommerville (1998); Pohl (2008).
[394] Jones (1996); Zowghi and Coulin (2005).
[395] OGC (2011).
[396] Neill and Laplante (2003); Zowghi and Coulin (2005).
[397] Zowghi and Coulin (2005).
[398] See for instance Glass (2002); Kotonya and Sommerville (1998); Macaulay (1996); Maiden and Rugg (1996); Wieringa, Maiden, Mead and Rolland (2005); Yadav, Bravoco, Chatfield and Rajkumar (1988).
[399] Constantine and Lockwood (2002).
[400] See section 3.3.
[401] Escalona and Koch (2003).
[402] Hallerstede and Bullinger (2010).

3.6.2   Design phase

In the design phase, requirements are translated into specifications of the software. In traditional software development, this phase focuses on the design of the software's and environment's architecture as they influence the structure and content of both, the software and its environment[403]. Under this aspect, architectural design is defined as "(1) The process of defining a collection of hardware and software components and their interfaces to establish the framework for the development of a computer system. [...] (2) The result of the process in (1)"[404]. When designing the software architecture, architectural patterns that describe best-practices should be followed[405]. The architectural design is mainly influenced by non-functional requirements[406]. If for instance multiple graphical interface designs are required, a model view controller architecture should be chosen[407]. The aim of a sound architecture is to improve performance, robustness, distributability and maintainability of software[408]. Apart from the architecture, which is invisible to the user[409], visible elements of the software have to be layouted[410].

As outlined in section 1.4, the social subsystem of an OIP can be influenced indirectly via the technical subsystem. Thus, the specification of features and their influence on social mechanisms has to be focused on. It is a non-trivial task to translate a requirement like "The OIP implements a point-based incentive system" into a feature, as the detailed design of it has to be mapped-out. For instance, the following questions have to be settled: "What does the innovator get points for?" and "How many?" and "What is the benefit of points?". The example shows that in OIP projects the formulation of a requirement can be generic and that an important task in the design phase is to interpret the requirements into a meaningful socio-technical system that facilitates desired mechanisms. The introduced design elements of OIPs[411] outline options for the most important decisions that have to be taken when designing an OIP. There is no blueprint for an OIP design, but only options. Different

---

[403] OGC (2011). In the wording of ITIL, the environment is also referred to as the operational model.
[404] IEEE (1990), p. 10.
[405] Bass et al. (2003); Bell (2009); Crumlish and Malone (2009); Fowler (2002); Garlan and Shaw (1993).
[406] Bosch (2000).
[407] Bell (2009); Krasner and Pope (1988).
[408] Bosch (2000).
[409] Like for instance the source code structure, the use of frameworks.
[410] Like for instance the interface structure. Bell (2009); Crumlish and Malone (2009); Galitz (2007); Mayhew (1991).
[411] See section 1.2.

combinations of design elements can lead to success even though their combinations might result for instance in a completely different motivational structure[412]. In the end, the OIP design has to complement the innovators' voluntary contributions[413] in order to foster a desired output[414] for the organizer.

3.6.3   Build phase

In the build phase the software and its architecture are implemented. Implementation is thereby defined as "(1) The process of translating a design into hardware components, software components, or both. [...] (2) The result of the process in (1)"[415]. Software components are purchased or built and subsequently integrated and tested. Testing takes place in the build *and* the deploy phase. In the build phase, testing verifies that functionality and manageability meet the specifications from the design phase. In this stage, pilot versions might be applied for testing[416]. Depending on the access to the target group, testing can include end users.

Three aspects are particularly important in this phase. Firstly, due to the availability of standard components, like for instance in open source software, *reuse* of existing software components is recommendable[417]. Software can be reused on different levels, ranging from the reuse of an abstract software design pattern, an object or a component to the reuse of a complete system. Reusing yields the benefits of higher development speed, lower risks and lower costs. If reusing existing components is not possible, new implementations should be designed to facilitate later reusability. Secondly, different version of a software should be tracked in a *configuration management* system. This system takes care of version management, system integration and problem tracking[418]. Configuration management systems ease collaboration of multiple developers and allow to roll back to old versions in case of a problem. The third aspect is the distinction between a development and live *environment*. Developments should be done in an environment separated from the

---

[412] See Hallerstede et al. (2010); Piller and Walcher (2006).

[413] Harhoff (2003).

[414] Borst (2010).

[415] IEEE (1990), p. 38. The implementation process is also referred to as coding. The result of building is a version of the software.

[416] OGC (2011).

[417] Sommerville (2011).

[418] Bellagio and Milligan (2005); Conradi and Westfechtel (1998); Hass and Hass (2003); Keyes (2004); Loeliger and McCullough (2012); Sommerville (2011).

live environment. This is particularly important for continuously evolving software in order to maintain a working live system[419].

For OIPs one task in this phase is the creation of the graphical layout and design. Whereas the graphical layout refers to the arrangement of components in an user interface, the graphical design refers to the visual design which uses graphics, colors etc. and thereby transports the aesthetics of a user interface[420]. Both depend on the purpose of the OIP. An expert-based OIP for instance, will potentially follow a serious graphical design, whereas a community-based OIP will tend to look playful[421].

3.6.4   Deploy phase

Once the system is built, deployment or the "release of a system or component to its customer or intended user"[422] starts. Therefore, the (changed) architecture has to be implemented to existing systems and the software has to be installed. Software version management tools help to deploy new components or a whole software[423]. They have to take care of version, configuration and release control[424]. After deployment is completed, tests have to be performed in order to verify the software in the new environment[425]. The process ends with a go live of the software, which marks the point at which users can start using it.

For OIPs, the deployment process has to be very efficient as deployment, owing to a potential perpetual beta paradigm, might be processed frequently[426]. For instance, some open innovation intermediaries offer standard OIPs as a software-as-a-service (SaaS) version to their customers[427]. These OIPs might require a lot of small releases to fix bugs and to add new functionality in order to incrementally the OIP[428]. On the other hand, there are custom OIPs, which, for example, might run only a short period of time[429]. For them, potentially fewer deployments are required, as a full

---

[419] Hoyer et al. (2007).

[420] See Fowler and Stanwick (2004); Stone, Jarrett, Woodroffe and Minocha (2005); Taylor, McWilliam, Forsyth and Wade (2002); Tidwell (2005).

[421] Hallerstede and Bullinger (2010).

[422] IEEE (1990), p. 25. Deployment is used synonymously to the IEEE notion of delivery.

[423] Conradi and Westfechtel (1998); Loeliger and McCullough (2012).

[424] OGC (2011).

[425] Dearle (2007); OGC (2011).

[426] See Fraternali (1999); Hoyer et al. (2007); Pressman (1998).

[427] For instance innocentive.com or unserAller.de.

[428] See Jeppesen and Lakhani (2010); Luethje and Herstatt (2004); Piller and Walcher (2006); Powell, Piccoli and Ives (2004).

[429] For instance smellfighters.com.

version of the OIP needs to be online right from the start. In addition to the traditional perception of the deploy phase that ends with the go live of software, marketing has to be conducted for an OIP in order attract relevant innovators[430].

3.6.5   Operate phase

During operation, support has to be given to the users and changes in the requirements have to be noticed. In addition, performance of the software is tracked[431]. Operation is thereby defined as the "period of time in the software life cycle during which a software product is employed in its operational environment, monitored for satisfactory performance, and modified as necessary to correct problems or to respond to changing requirements"[432].

Tasks in this phase vary from its traditional interpretation because a different focus has to be taken for OIPs than for standard software. OIPs have to be managed actively in order to enable their potential use[433]. Therefore, not only questions from users, i.e. the innovators, have to be answered, but innovators have to be motivated proactively as well[434]. Another challenge in this phase is to identify bugs, as the users in a web context have low barriers to leave the OIP in case of a problem[435]. In addition, *intellectual property rights* have to be managed actively and transferred to the organizer [436].

3.6.6   Optimize phase

The final ALM phase is optimize. During this phase the results from operations and changes in the environment are analyzed and reacted upon. Based on the analysis, required changes to the software are identified and initiated. The aim is to maintain or improve the software quality or to lower cost by fault repairs[437], improving functionality or adding functionality[438]. According to Erlikh[439] 85-90% of software

---

[430] Adamczyk et al. (2010), (2012); Terwiesch and Xu (2008).

[431] OGC (2002).

[432] IEEE (1990), p. 52.

[433] Adamczyk et al. (2011); Moon and Sproull (2008).

[434] See for instance Adamczyk et al. (2010); Antikainen and Vaataja (2010); Brabham (2009); Fueller et al. (2009).

[435] Bell (2009); Hallerstede et al. (2011); Whitworth, Fjermestad and Mahinda (2006).

[436] Chesbrough (2006a); Huizingh (2011); Terwiesch and Xu (2008).

[437] Fault repairs is used synonymously to the term bug fixing.

[438] OGC (2011); Sommerville (2011).

[439] Erlikh (2000).

costs are operation and optimization costs in large systems[440]. During these maintenance activities, more budget is spent to implement new requirements than to repair faults[441]. Thus, the software should be designed to complement functional additions right from the beginning. Figure 18 shows the estimated share of fault repairs, environmental adaptions and functional additions/ modifications.

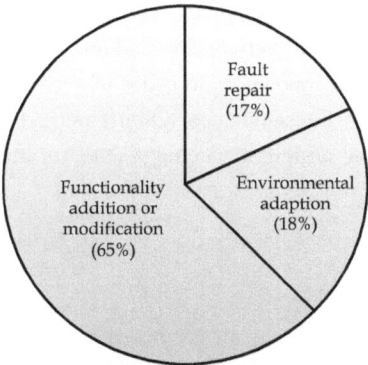

Figure 18: Distribution of maintenance effort in software projects[442]

In order to identify potential improvements, web metrics should be applied due to the web-based character of OIPs. In addition to user feedback, server logs can be analyzed and web analytics can be used[443]. In an OIP environment, continuous improvement according to a perpetual-beta paradigm might apply[444]. Chapter 4 summarizes the findings of this part concerning open innovation platforms, open innovation intermediaries and the lifecycle model for OIPs.

---

[440] Although OIP projects are not considered large (cf. Diener and Piller 2010), the tendency of relevant maintenance cost is considered the same.
[441] Davidsen and Krogstie (2010); Nosek and Palvia (1990); Sousa (1998).
[442] Sommerville (2011), p. 244.
[443] Elbaum, Rothermel, Karre and Fisher II (2005); Weischedel and Huizingh (2006).
[444] See Hoyer et al. (2007).

4 Summary of Part II

This part set the foundations for the analyses to come. Open innovation platforms (OIPs) were introduced as socio-technical systems that incorporate IT-based tools for open innovation in a virtual environment. These tools include *innovation contests, innovation communities, innovation market places* and *innovation toolkits*. The aim of OIPs is to connect organizers and innovators in order to solve an organizer's innovation problem. Open innovation intermediaries (OIIs) emerged as professional lifecycle managers of OIPs, i.e. they design and manage OIPs for an organizer, as shown in Figure 19.

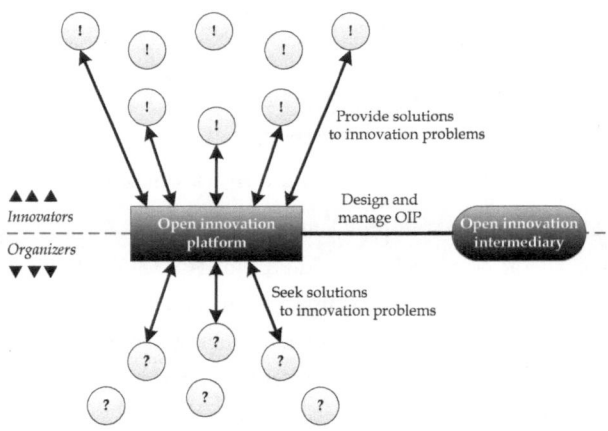

Figure 19: Open innovation intermediary as OIP lifecycle manager

Multiple challenges arise due to the virtual environment of OIPs. These are *selecting the right problems, formulating problems, employees' reluctance,* as well as *facilitating software-mediated knowledge transfer*. Open innovation intermediaries fulfill multiple functions to support the *connection and collaboration* of organizers and innovators by providing *technological services and support*. It was shown that open innovation intermediaries lack an informed approach to design and manage OIPs, which is vital for an OIP's success. Thus, application lifecycle management with its six phases was selected as a suitable *lifecycle model* to structure the design and management of OIPs. The implications of the particular field are considered in an adaption of ALM to it, namely the OIP lifecycle model (OIP-LM), which is shown in Figure 20. An OIP lifecycle manager, such as an open innovation intermediary, has to consider both, social and technical components of an OIP's socio-technical system, as well as their interdependence in each phase of the OIP-LM.

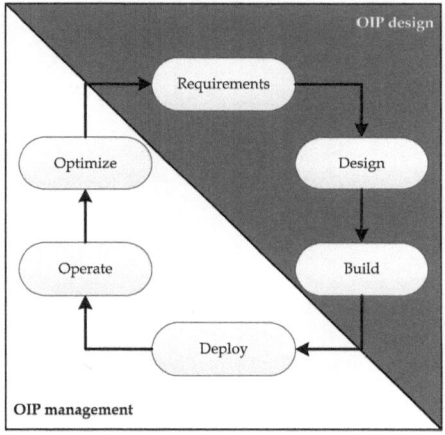

Figure 20: The OIP lifecycle model

Table 8 summarizes the activities in each phase of the OIP lifecycle model.

Table 8: Activities in the six phases of the OIP lifecycle model

Stage	Phase	Activities
OIP design	Requirements	Gathering of functional, manageability, usability, architectural, interface, service level and non-functional requirements
	Design	Translating requirements into specifications of the OIP's architecture and interfaces while considering the social *and* the technical subsystem as well as software design patterns
	Build	Implementing and testing the OIP's design with special regards to functionality and graphical design while taking a make or buy decision for each component
OIP management	Deploy	Installation of the OIP, rollout to the users and marketing to attract innovators
	Operate	Keeping the OIP running, identifying required changes, providing support to users and active community management
	Optimize	Analyzing the OIP and feedback based on operations in order to maintain or improve the OIP quality or to lower cost

Part III

Empirical study

1 Case study design

Part II set the foundations for the three facets of OIP lifecycle management, namely OIIs, OIPs and a model for OIP lifecycle management. The present *Part III* introduces the empirical data of this work that serve to analyze the three facets. As elaborated in the research design[445], case study research is chosen as an explorative research approach[446] in order to better understand the lifecycle of OIPs and its management. To outline the research approach, firstly, the research method, namely the case study design, is described (*chapter 1*). Secondly, findings from three cases of OIP lifecycle management by OIIs are portrayed. The cases are structured according to the three facets of OIP lifecycle management: The OII, its OIP and its lifecycle management. The individual ways of the OIIs to manage an OIP's lifecycle are highlighted (*chapters 2 to 4*). The part closes with a summary of findings from the three cases (*chapter 5*).

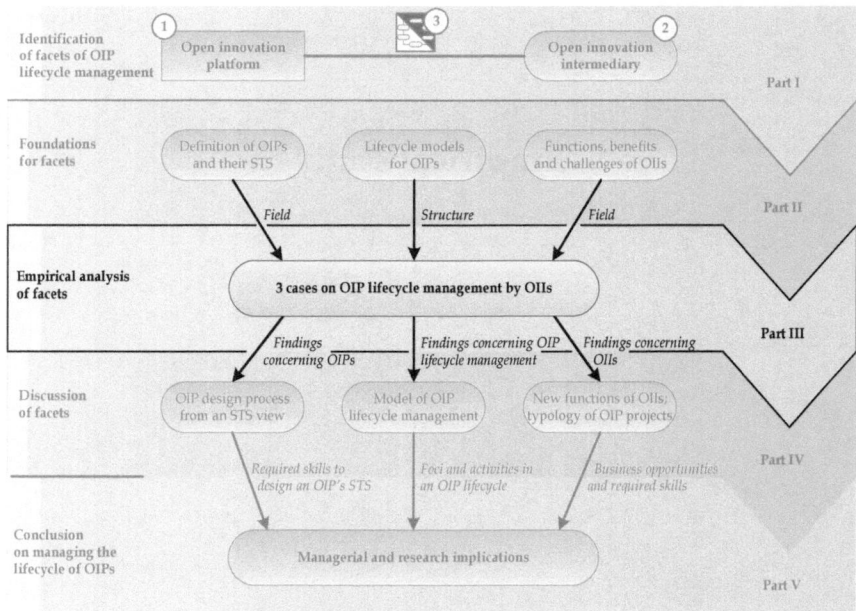

Figure 21: Current progress in the research design

---

[445] See Part I.3 and Figure 21.
[446] Robson (2002); Schnell et al. (2008).

This first chapter introduces the research method applied in this thesis. The research approach follows a multiple holistic case study design according to Yin[447]. The structure and content of this chapter and the case study mainly follow the guidelines for case studies developed by Runeson and Hoest[448] as well as the quality standards formulated by Dubé and Paré[449], Creswell[450], and Robson[451].

The first *section 1.1* gives reasons for the choice of a case study design. Subsequently, the case selection (*section 1.2*) and data collection procedures (*section 1.3*) are described. They are followed by the data analysis (*section 1.4*) and validity procedures (*section 1.5*).

1.1 Case study as a research method

Managing the lifecycle of OIPs is closely related to the context of an OIP project and thus subject to unknown factors that influence OIP lifecycle management. In this setting, case study research is the method of choice to explore OIP lifecycle management and to derive corresponding guidelines, as case studies can assess a phenomenon within its context and reveal underlying relationships by generating rich data[452]. Additionally, the research questions highlighted in Part I.2 aim at the "who" and "how" of OIP lifecycle management which calls for a case study approach as well[453].

To identify the relationships already mentioned and to elaborate on the OIP lifecycle model[454] and its influencing factors, in-depth case analyses of OIP projects are conducted and subsequently compared applying pattern-matching in a cross-case analysis[455]. In this case study, a case is constituted by an *open innovation intermediary*. Each OII represents a case with a single unit of analysis, namely *the way of designing and managing OIPs*. According to this definition, a case reflects an integrated view on OIP lifecycle management by a particular OII, mainly building on a single or multiple OIP projects the open innovation intermediary ran. The cases draw on multiple data sources like interviews, documents and websites. The context of a case includes, but is not limited to, the context of the OII (e.g. legal constraints), the organizer of the OIP

---

[447] Yin (2009).
[448] Runeson and Hoest (2008).
[449] Dubé and Paré (2003).
[450] Creswell (2007).
[451] Robson (2002).
[452] Benbasat et al. (1987); Borchardt and Goethlich (2007); Creswell (2007); Yin (2009).
[453] Yin (2009).
[454] See Part II.4.
[455] Yin (2009).

(i.e. the OII's client), the OIP's target group (i.e. the innovators) and the aim of the OIP. The multiple holistic cases are compared to literally and theoretically replicate results across cases[456]. Data analysis and reporting structure follows the guidelines for linear-analytic case studies as provided by Dubé and Paré[457], Runeson and Hoest[458] as well as Yin[459]. The present chapter on the case study's design is particularly inspired by the foremost authors, who stress the importance of details in the context of data collection and analysis. Although data collection is conducted exploratively, data description and analysis is informed by the foundation of Part II. The findings are structured along the three facets of OIP lifecycle management. This approach thus follows Yin's[460] liberal concept of a case study, which, in contrast to Stake's[461] conservative concept, allows for theoretical input to frame a case study. The case selection strategy is introduced next.

1.2   Case selection

Cases should be selected in order to provide insights on good-practices concerning OIP lifecycle management. Due to their specialization in the field of OIP lifecycle management and many OIP projects processed[462], open innovation intermediaries have potentially more experience in designing and managing OIPs than other OIP lifecycle managers, like for instance companies that run single OIP projects. Consequently, cases should be drawn from *open innovation intermediaries*[463].

In their study on the market of open innovation, Diener and Piller[464] investigated open innovation intermediaries[465]. They aimed at comprehensively covering the market for open innovation, hence identifying as many OIIs as possible. They "decided for a very low threshold of being classified as an OIA [OII], because getting a large variability in the data set supported our objective of mapping the open

---

[456] Yin (2009).

[457] Dubé and Paré (2003).

[458] Runeson and Hoest (2008).

[459] Yin (2009).

[460] Yin (2009).

[461] Stake (1995).

[462] See Diener and Piller (2010).

[463] For a definition of OIIs see Part II.2. See also Part I.3.

[464] Diener and Piller (2010).

[465] They thereby cover open innovation intermediaries, i.e. those intermediaries using OIPs, and other intermediaries in the field of open innovation, i.e. those intermediaries *not* using OIPs to bridge the gap between organizers and innovators (see the definition of OII in Part II.2). The latter ones are excluded from the list in filter I), as set out below.

innovation landscape"[466]. Thus, as the study provides a comprehensive list of OIIs in the market, it provides a good starting point to select OIIs for the case study.

Diener and Piller[467] identified 43 OIIs. Data collection took place between October 2007 and June 2009[468]. In addition to that, the authors note eleven OIIs that came to their attention only after data collection for the study had been completed. Thus, there are 54 OIIs that might serve the case study. 39 of them are still active in the market and can be investigated. They are listed in Table 9.

Four filters are applied to the OIIs in the market in order to select appropriate ones for the case study[469]. In *filter I)*, open innovation intermediaries have to *use IT-based tools for open innovation*, i.e. *innovation contests, innovation communities, innovation marketplaces* or *innovation toolkits* in order to be eligible for this study. 31 of the 39 OIIs remain after applying this filter and could therefore be researched.

In *filter II)*, cultural influences, such as the preference of different software lifecycle models in America and Europe[470], are excluded from the analyses in order to make the cases comparable. Thus, all OIIs in the case study should be based and offer services in *Western Europe*[471]. Applying this filter, eight OIIs remain.

In *filter III)*, *access to the OIIs* is a selection mechanism to ensure that the companies are willing to share confidential information. Although this access-based selection mechanism might lead to a biased case selection, I argue for the benefits of the access-based selection (e.g. access to confidential information), which outweighs the drawbacks (e.g. biased case selection) in this explorative setting[472]. Two out of the eight remaining OIIs[473] did not agree to participate in the study of Diener and Piller[474]. Thus, it is assumed that they are not willing to share information, especially confidential information, for the intended case study. Consequently, they are excluded, resulting in six *qualifying OIIs* for the study, namely Atizo, Brainfloor, Hype, HYVE, innosabi and VOdA.

---

[466] Diener and Piller (2010), p. 32.
[467] Diener and Piller (2010).
[468] As the OIIs for the case study should be established in the market and have some experience in managing the lifecycle of OIPs, it is *not* necessary to update the assessment of Diener and Piller (2010) in order to identify OIIs that are founded recently, as these would not yet be established.
[469] See Table 9.
[470] See for instance Carmel, Whitaker and George (1993); Carmel (1999); Krishna, Sahay and Walsham (2004).
[471] This also eases access to the OIIs.
[472] See Andrews and Pradhan (2001).
[473] 99Designs and Wilogo. Legal forms of the OIIs are omitted to ease readability.
[474] Diener and Piller (2010).

Table 9: Selection of OIIs for the case study[475]

OII (legal form omitted)	Homepage	Innovation contest	Innovation community	Innovation market place	Innovation toolkit	Western Europe	Access	Selection
Cassiber	www.cassiber.com					✓	✓	
Elephant Design	www.elephant-design.com						✓	
Elephant Design + Strategy	www.elephantdesign.com						✓	
Future Lab Consulting	www.futurelab.de			*Filter I)*				
Gen 3 Partners	www.gen3partners.com							
Invention Machine	www.inventionmachine.com						✓	
LEAD Innovation Mgmt	www.lead-innovation.com					✓		
Verhaert	www.verhaert.com							
Battle of Concepts	www.battleofconcepts.nl	✓					✓	
Big Idea Group	www.bigideagroup.net	✓	✓				✓	
Brain Reactions	www.brainreactions.com	✓					✓	
Communispace	www.communispace.com		✓				✓	
Crowdspring	www.crowdspring.com	✓	✓				✓	
Elance	www.elance.com		✓	✓			✓	
Fla Fabric	www.favelafabric.com		✓				✓	
Fronteer	www.fronteerstrategy.com				✓		✓	
Guru	www.guru.com		✓	✓			✓	
IBM	www.collaborationjam.com	✓					✓	
Idea Connection	www.ideaconnection.com	✓	✓		✓		✓	
Ideas To Go	www.ideastogo.com	✓				*II)*	✓	
Innocentive	www.innocentive.com	✓	✓	✓			✓	
Innovation Framework	www.innovation-framework.com	✓	✓		✓		✓	
inpama (formerly Ideawicket)	www.inpama.com	✓	✓		✓		✓	
NineSigma	www.ninesigma.com	✓	✓				✓	
Skild (formerly Idea Crossing)	www.skild.com	✓	✓				✓	
Spigit	www.spigit.com	✓					✓	
Sitepoint	www.sitepoint.com	✓						
Venture2	www.venture2.net		✓		✓		✓	
vWorker (formerly Rent-a-coder)	www.vworker.com	✓	✓				✓	
Yet2.com	www.yet2.com	✓		✓			✓	
Your Encore	www.yourencore.com	✓	✓		✓		✓	
99 Designs	www.99designs.de	✓				✓	*III)*	
Wilogo Not qualifying OIIs	www.wilogo.com	✓	✓			✓		
Brainfloor Qualifying OIIs	www.brainfloor.com	✓				✓	✓	
Hype	www.hypeinnovation.com		✓			✓	✓	*IV)*
VOdA	www.vo-agentur.de	✓	✓		✓	✓	✓	
Atizo	www.atizo.com	✓	✓	✓		✓	✓	✓
HYVE	www.hyve.de	✓	✓		✓	✓	✓	✓
innosabi	www.innosabi.com		✓		✓	✓	✓	✓

---

[475] Based on Diener and Piller (2010).

Finally, in *filter IV)*, cases have to be selected from the six qualifying OIIs. The selected OIIs should cover OIPs that contain *innovation contests, innovation communities, innovation marketplaces* as well as *innovation toolkits* in order to ensure generalizability of the results for all OIPs. Atizo is the only qualifying OII that offers innovation market places. Thus, *Atizo* is selected for the case study. According to Diener and Piller[476] and the OIIs' homepages, the OIIs Hype, HYVE and VOdA resemble each other in terms of services they offer and their organizational profile[477]. HYVE is one of the first movers, most established and market leading consultancies for open innovation in Germany[478]. In addition to that, the author has access to HYVE[479]. Consequently, *HYVE* is selected as a second case. Two OIIs remain, namely Brainfloor and innosabi. While Brainfloor offers innovation contests, innosabi offers an innovation community and innovation toolkits. As both of the previously selected OIIs already cover innovation contests and only one offers innovation toolkits, innosabi is selected as the third case in order to have an even representation of IT-based tools for open innovation in the case study[480]. Also, access to innosabi is granted.

This information oriented selection strategy[481] aims at a maximum variation among the researched cases[482] while focusing on a professional, informed approach to OIP lifecycle management. This allows identifying relevant influencing factors in OIP projects that are not related to cultural differences. The three cases selected for this case study are shown in Table 10.

Table 10: Overview of cases included in the case study

Case name	OII	OIP	Chapter
innosabi	innosabi GmbH	unserAller.de	2
HYVE	HYVE AG	HYVE IdeaNet©	3
Atizo	Atizo AG	Atizo.com	4

While the *innosabi case* incorporates an innovation community and applies innovation toolkits, the *HYVE case* covers innovation contests, innovation communities and

---

[476] Diener and Piller (2010).
[477] For instance number of employees, year of foundation, all are agencies stemming from open innovation.
[478] Williams (2011).
[479] See the argumentation concerning *access to the OII* above.
[480] See also Table 11.
[481] See Flyvbjerg (2006).
[482] See Carmel (1999); Verworn and Herstatt (2000).

innovation toolkits. Finally, the *Atizo case* comprises an innovation community, innovation contests and some traits of an innovation marketplace. The coverage of IT-based tools for open innovation by the selected open innovation intermediaries is shown in Table 11. The following outlines the data collection procedures applied to assess the cases.

Table 11: Coverage of IT-based tools for open innovation by case

IT-based tool for open innovation	innosabi	HYVE	Atizo
Innovation contest		✓	✓
Innovation community	✓	✓	✓
Innovation market place			✓
Innovation toolkit	✓	✓	

1.3   Data collection procedures

Data collection took place from April 2011 until August 2012. The main data sources for the case elicitations are semi-structured, explorative interviews[483] with executives and other employees of the open innovation intermediaries. Interview partners were selected to represent all relevant corporate actors to an OIP project including executives, project managers, community managers and developers. This way, the interview partners represented similar functions in each case, including at least the before mentioned ones, in order to improve comparability of data. Executives were interviewed first, as they tend to have the best overview of the organization's structure. From there on, a pyramiding approach[484] was chosen to identify additional interview partners and other relevant data sources like documents. Data sources, especially interview partners, were added until saturation was reached and no new information had to be expected from additional ones[485]. Topics of the interviews included the company in general, its competencies and processes, the concept and lifecycle management of their OIP as well as the division of labor between the OII, the innovators and the organizer. At the end of each interview, interviewees were confronted with an open question that allowed them to highlight whatever they considered relevant concerning designing and managing OIPs. According to the

---

[483] Robson (2002).
[484] Von Hippel et al. (2009).
[485] Corbin and Strauss (2008).

flexible research design of case studies, the interview guideline was constantly adapted to new findings[486]. It stabilized during the elicitation of the second case.

The interviews were conducted face-to-face in the OIIs' offices[487] using the critical incidents technique[488] while following the interview technique guidelines of Robson[489]. In alignment with the critical incidents technique, interviewees were asked to draw on extreme examples of successful and unsuccessful OIP lifecycle management. In addition to that, the interviewer pushed on extreme examples and tried to identify influencing factors. The interview sessions were taped and transcribed. The interview partners volunteered to participate in the study and had their CEO's permission to reveal confidential information. At the beginning of an interview, the interview partners were informed about the aim and method of the study. The interviewer built up a casual and conversational atmosphere of trust[490]. The interviewees had the opportunity to review the transcripts before making an approval. This process was put in place in order to grant the protection of confidential information. By means of this option, the interviewees were able to speak freely during the interview. One interview partner used the opportunity to review the interview prior to its approval. However, no relevant information was excluded. Publication of any information collected in the interviews or from internal documents is subject to prior approval by the concerned OII. Table 12 gives an overview of the included interviews.

In addition to the interview data, further material was used for the case study. Firstly, the OIIs' online presences were scanned for public information. Secondly, internal documents (strategy presentations, visual material, documentation etc.) were requested from the interview partners. Thirdly, the OIPs themselves were analyzed. Therefore, active participation in different innovation projects that were running on the OIPs was sought. In order to explore for instance backend capabilities, logins to live or development environments were acquired. By this, all relevant views on an OIP, i.e. representing all stakeholders in such a system like innovators, community managers and administrators, were covered. Fourthly, material that generates an overview of an OII and its activities was compiled. This includes for instance lists of employees and their tasks, organizational charts or project lists. Finally, existing

---

[486] See Runeson and Hoest (2008).
[487] The interviews with Atizo were conducted remotely using Skype video conferencing.
[488] Flanagan (1954).
[489] Robson (2002).
[490] See Andrews and Pradhan (2001).

research concerning the OIIs, like prior case studies, were included. The following section introduces how the collected data was analyzed.

Table 12: Interviews included in the case study

Reference	Case	Position of interviewee	Duration [min]	Major insights
I_CEO	innosabi	CEO & Founder	177	Strategic positioning and outlook; development of the company and its OIP; community management
I_CIO	innosabi	CIO & Founder	81	IT organization; design of the OIP; origin and prioritization of requirements
I_CTO	innosabi	CTO & Founder	98	Architecture and versions of the OIP; division of tasks
H_CEO	HYVE	CEO & Founder	76	Development of the company; strategic positioning
H_CPM	HYVE	CEO & Senior Project Manager	62	OIP versions and lifecycles; strategy HYVE IdeaNet©; overview of projects
H_SPM1	HYVE	Senior Project Manager	75	Relationship to organizers
H_SPM2	HYVE	Senior Project Manager	69	Deep insights into single OIP project
H_PM1	HYVE	Project Manager	67	Deep insights into single OIP project
H_PM2	HYVE	Project Manager	54	Role of the project manager
H_DEV	HYVE	Senior Developer	68	Roles and interfaces of the developer
A_CTO	Atizo	CTO & Founder	56	Overview of the company; technical aspects; operations
A_CEO	Atizo	CEO & Founder	71	Moderation of workshops; community management; division of tasks
A_ADM	Atizo	Project administrator	57	Innovation project setup and management; internal processes

1.4 Analysis procedures

Data preparation and analysis was split into *five steps* following Mayring's [491] approach to qualitative content analysis. Data analysis (i.e. especially coding of the interviews and comparing, sorting and integrating codes) was processed using

---

[491] Mayring (2008).

MAXQDA 10[492]. Figure 22 depicts the process of data analysis. It is further elaborated in the following paragraphs.

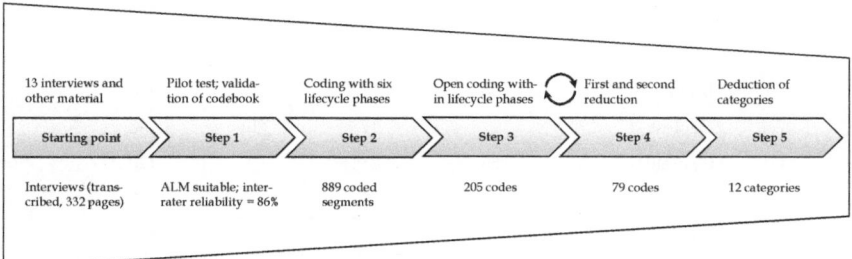

Figure 22: Process of data analysis

Firstly, in a *pilot test*[493], selected transcripts of the innosabi case, that represented approximately 15% of the data, were coded according to the six phases of the OIP-LM[494] by two coders. In the course of this, the codebook was defined and refined, resulting in an inter-rater reliability of 86%. The average inter-rater reliability results from difficulties in distinguishing descriptions of requirements and functions, which are either assigned to the requirements phase or the design phase of the OIP-LM[495]. Ignoring this fuzziness – which is possible as it does not affect the results – inter-rater reliability rises to 98%. The pilot test has shown that the OIP-LM with its six phases is suitable for structuring the activities during the design and management of OIPs.

In a *second step*, transcripts from all interviews[496] were *coded* according to the lifecycle phases *deduced from the OIP-LM* to structure the data. Table 13 shows the amount of coded segments[497] per case and lifecycle phase. The decreasing number of coded segments per case reflects the increasing focus of the study in the course of the continuous iterations of data collection and data analysis[498].

---

[492] Lindsay (2004).

[493] Opposing to a pre-test that tests (parts) of a study in a laboratory setting, a pilot test processes the whole study including a potential analysis in a field setting. Consequently, the data from the pilot test can be used for the main study. Yin (2009).

[494] See Part II.3.6.

[495] This issue is explicitly addressed in the codebook.

[496] The interviews summed up to 332 pages, which is equal to 139.589 words or 901.143 characters (including spaces).

[497] A coded segment is a text passage that is assigned to a code. A coded segment might comprise a sentence, a paragraph or multiple paragraphs.

[498] See Yin (2009).

Table 13: Coded segments per case and lifecycle phase

Phase	innosabi	HYVE	Atizo	Total
Requirements	106	84	22	212
Design	112	118	58	288
Build	58	77	11	146
Deploy	22	12	14	48
Operate	50	21	49	120
Optimize	59	7	9	75
Total	407	319	163	889

The *third step* incorporated inductive *open coding* of the coded segments to explore activities, influencing factors and patterns *within each phase* of the OIP-LM. New aspects of OIP lifecycle management were initially coded as concepts and matched by constant comparison[499].

In *step four*, after having coded all interviews, codes were *compared, sorted and integrated* in order to reduce them to a higher level of abstraction. In the design phase for instance, the codes "reusing existing source code" and "modularity" (step three) were integrated into "modular architecture" (step four).

In a *fifth step*, *categories* are derived from the codes in order to draw conclusions[500]. Table 14 summarizes the results of open coding, reduction and the amount of categories derived.

In addition to the interview data, further material was used for the case study, including the OIIs' online presences, public and internal documents (strategy presentations, visual material, documentation etc.) as well as the OIPs. The different data sources were triangulated[501].

Subsequent to the within-case analyses, which are introduced in chapters 2 to 4, the findings will be discussed in a cross-case analysis in Part IV[502]. In order to ensure validity of results, the next section addresses threads to the validity of the study and applied strategies to cope with these threads.

---

[499] Strauss and Corbin (1998).
[500] Mayring (2008).
[501] Yin (2009).
[502] Miles and Huberman (1994); Yin (2009).

Table 14: Number of codes and categories per OIP-LM phase

Phase	Open coding	Reduction	Categories
Requirements	40	20	5
Design	65	21	4
Build	39	9	4
Deploy	10	9	3
Operate	25	10	4
Optimize	26	10	4
Total	205	79	24

1.5   Validity procedures

In accordance with Yin's[503] classification scheme of a case study's validity, the following four aspects were considered to improve the research quality[504]. Despite the use of a structured method as introduced in the section above, additional strategies to deal with threads to validity draw on Robson[505] as well as Padgett[506].

Construct validity and corresponding description issues were addressed by debriefings and feedback loops among the researchers (peer-debriefing) and interviewees (member checking)[507]. Preliminary results for instance, were discussed with the interviewees and among peer researchers in order to ensure a common understanding and appropriate interpretation of observed phenomena[508]. Additionally, the pyramiding approach to identify data sources[509] helped to collect a rich set of information. A prolonged involvement in the OII tackled reactivity and a respondent bias and increased access to valid data.

Internal validity and interpretational concerns are addressed by data and observer triangulation to reduce the risk of false interpretations[510]. This was backed

---

[503] Yin (2009).
[504] See the comments on critical realism in Part I.3.
[505] Robson (2002).
[506] Padgett (1998).
[507] See Maxwell (1992).
[508] See also Part V.4.
[509] See section 1.3.
[510] See Denzin (2009); Maxwell (1992).

up by a pilot test to confirm the suitability of the OIP-LM for the study and the overall approach to analyze the data[511].

External validity is facilitated by the multiple case design and the case selection strategy which aims at a maximum variation to identify influencing factors[512]. By the application of a lifecycle model to structure the findings (i.e. the OIP-LM), transferability to other OIP projects and settings is facilitated[513].

Reliability [514] is addressed by the application and communication of a standardized research approach with a case study database that contains all relevant information (e.g. transcripts, documents), documentation (e.g. the case study protocol, an overview of the interviews, timelines) and codebooks. The research process includes observer triangulation to additionally reduce a researcher's bias[515]. Table 15 summarizes the case study design. The following chapter introduces the first case of this case study.

Table 15: Summary of case study design

Case study design	Multiple, in-depth, holistic, explorative, linear-analytic case study
	Case selection strategy: informed, aiming at maximum variation
	Data sources: interviews, internal and public documents, websites
	Case description and analysis structured according to the OIP-LM
	Open coding within OIP-LM phases
	Pattern-matching; literal and theoretical replication
	Construct, internal and external validity, as well as reliability procedures apply

---

[511] See section 1.3.
[512] See section 1.2.
[513] See Part I.3.
[514] According to Yin (2009) and in contrast to the notion in quantitative analyses (Schnell et al. 2008), reliability is subsumed under a case study's validity.
[515] See Denzin (2009).

2 Case 1: innosabi

The first case investigates the open innovation intermediary innosabi[516] with its OIP unserAller.de. Table 16 summarizes the key facts on the innosabi case.

Table 16: Key facts on the innosabi case

OII	innosabi GmbH (referenced as innosabi)
Description	innovation agency with a standardized OIP that runs multiple innovation projects on the same platform and in a default graphical design
Foundation	2007
Employees	10 (including 4 founders)
Homepage	www.innosabi.com
OIP	www.unserAller.de
Tools	innovation community; innovation toolkit

This case and the two subsequent cases in *chapters* 3 and 4 follow a common structure induced by the three facets of OIP lifecycle management identified in *Part I*: open innovation platforms, open innovation intermediaries and a model of OIP lifecycle management[517].

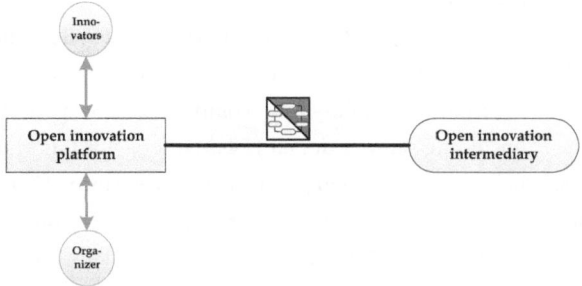

Figure 23: Three facets of OIP lifecycle management to structure the case analysis

---

[516] In the following, the legal form (e.g. GmbH) is omitted when referencing to an OII. This eases readability. If the case and not the OII is addressed, it is explicitly stated (e.g. 'in the innosabi case').
[517] See Figure 23.

Firstly, the open innovation intermediary as a company is briefly introduced (*section 2.1*). Secondly, the open innovation platform and its processes are outlined (*section 2.2*). Finally, the open innovation intermediary's approach to design and manage its OIP is elaborated along the six phases of the OIP lifecycle model, as shown in Figure 24 (*section 2.3*). Each case is briefly summarized at its end and conclusions are drawn from it (*section 2.4*).

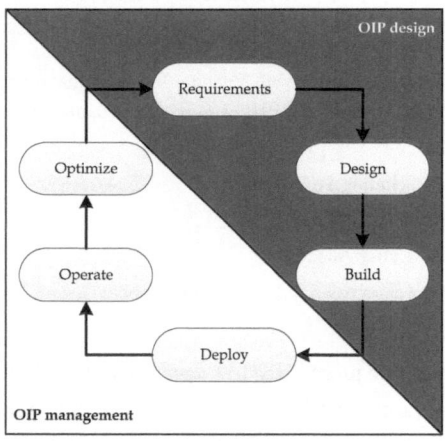

Figure 24: The OIP lifecycle model

2.1   The open innovation intermediary

innosabi sees itself as a service provider that offers technology for the integration of consumers into the innovation process of large and mid-sized companies. Their portfolio contains multiple methods, including self-developed IT-based tools for this aim. innosabi was founded in 2007 and has been growing constantly ever since. Their first activities included projects on pico-jobs and open innovation consulting. Both require intensive usage of human resources. In 2010, they added unserAller, an open innovation platform, to their portfolio. innosabi's clients, i.e. organizers, can run standardized innovation projects on unserAller which are managed by the OII, who in turn earns a software license and community management fee. Their clients include Daimler, Siemens, dm, Goertz, OSRAM, LEXA and other large and mid-sized producers of consumer goods. In June 2011, innosabi has been awarded the most successful information and communication technology-based start-up by Germany's federal ministry of economics and technology (BMWi). The OII was honored personally by federal minister Dr. Philipp Roesler[518].

innosabi is led by its four founders which are each responsible for a corporate function (CEO, CTO[519], CFO[520] and CMO[521]). The founders are supported by six employees who take care of specialized tasks like community management, programming or sales. Despite the founders' authority, hierarchies do not apply in the team. innosabi maintains a close and friendly relationship with organizers and tries to adapt to their needs as far as possible. innosabi lives a perpetual-beta approach. They try whatever they think is reasonable, maintaining things that work and discarding those which do not. Thus, the social as well as technical subsystem of their OIP is improved continuously. Beyond their OIP, this perpetual-beta paradigm applies to all activities and business areas including their methods, processes and tools. Today, innosabi's core asset and focus is the open innovation platform unserAller, where companies can develop products together with consumers. The OIP unserAller is introduced in the following section.

2.2   The open innovation platform

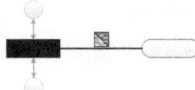

The following two subsections introduce the OIP unserAller (*subsection 2.2.1*) as well as its processes and core functionalities (*subsection 2.2.2*).

2.2.1   Overview

About 13,000 registered users, mostly from Germany, currently belong to the voluntary circle of innovators of the unserAller community, which started with 81 fans on Facebook in June 2010 (as of February 2012). The OIP unserAller[522] is specialized in the development of line extensions for producers of fast moving consumer goods.

Consumers can participate in the development of new products based on a standardized development process developed by innosabi. Projects run consecutively and are actively managed by innosabi's community managers. A list of major projects can be found in Table 17.

---

[518] BMWi (2011).
[519] Chief technical officer.
[520] Chief financial officer.
[521] Chief marketing officer.
[522] See Figure 25.

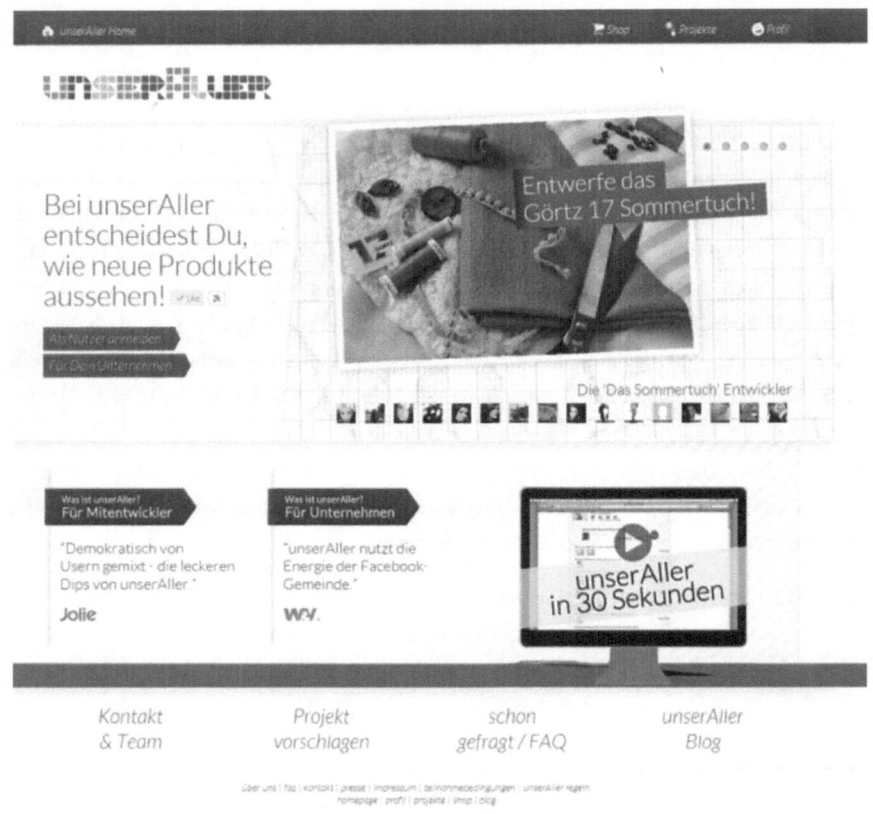

Figure 25: Homepage of unserAller

The declared aim of innosabi is the development of high quality products in the unserAller community that meet consumers' needs. To reach this aim, innosabi acts as an OII between the organizer and the organizers's consumers. innosabi adopts multiple roles: they divide the organizers' innovation problem into small tasks which can be handled by the community, they offer an OIP and processes in order to collaboratively work on the innovation task, and they are responsible for the management of the innovation project. Details of the task division between the organizer, community and OII are described below. In order to illustrate the processes, the project of the Balea shower gel represents an adequate example.

Table 17: Selected projects on unserAller (in chronological order)

Organizer	Project	Duration	Participants / Contributions / Comments	Result
Mari-Senf	Mustard dip	62 days	1.500	3 types of mustard: cassis-plum-mint, mango-honey and wasabi
innosabi	Bath salts	67 days	2.000 / 180	6 bath scoops for individual mixture: cocktail, milk, glitter, health, almond oil and milk-honey
Egi-Oel	Salad dressing	59 days	250 / 1.000	Country dressings: Spanish, Croatian and Thai dressing
dm-Drogerie-markt	Balea shower gel for the cold season	49 days	2.700 / 2.400 / 4.800	Balea shower gel „Eisschimmer"
anonymous	Milk-based and satiating snack	67 days		Healthy tapioca-oats-chocolate in sweet and salty variations
Restaurant Gesellschafts-raum	Chutney creation	26 days		Onion/peach-chutney with chili and balsamic vinegar
Goertz 17	Summer scarf	71 days		3 scarfs: Peacock feather, cherry blossom and butterfly in different materials

2.2.2 Processes and functionality

innosabi designed a five-stage development process to create new products collaboratively. It consists of the phases *reason, design, material, name* and *packaging,* as shown in Table 18.

Each of the five phases is dedicated to a part of the organizer's innovation problem (e.g. the "search for a shower gel for the cold season" in the Balea project). The individual arrangement of the phases, as well as the tasks within one phase, are elaborated in a workshop with the OII and the organizer. Phases may be omitted (like in the Balea project, see Table 18) or processed multiple times for instance in order to define the material of different parts of a product.

Table 18: Phases of the unserAller development process[523]

Phase	Aim	Sample task	Sample result
Reason	Determining the motto	What is our common motto for the new Balea shower gel?	Diamonds and Ice
Design	Determining the product's look	(Phase omitted in this project)	-
Material	Determining the product's material or composition	What scent does our shower gel have, does it shimmer and which color has it got?	Fruity, tasty smell, cream shower gel with shimmer, light turquoise color
Name	Finding a name	What is an adequate name for our shower gel?	Eisschimmer
Packaging	Type and design of packaging	How is the label of the shower gel designed?	Sketch of the label's design

The steps in each phase of the development process on unserAller are the same: It starts with a period of suggestions and discussions by the community in order to gather ideas and to further develop them, continues with a short review period for the organizer and ends with an evaluation period for the community. This setup aims at ensuring the strategic as well as the technical fit of community innovations to the organizer while at the same time maintaining a sense of ownership by the community. After the completion of a project, the organizer manufactures the new product and offers it at the unserAller shop and through the producers' usual distribution channels.

Overview of the course of one phase during the unserAller development process

The activity diagram in Figure 26 shows the course of one phase of innosabi's development process. Phases (e.g. reason, design, material) run consecutively. The outcome of a phase serves as input for the subsequent phases. A phase starts with the announcement of a task, which is developed by the OII. The following paragraphs describe the further process of a phase.

---

[523] Illustrated at the example of the Balea project.

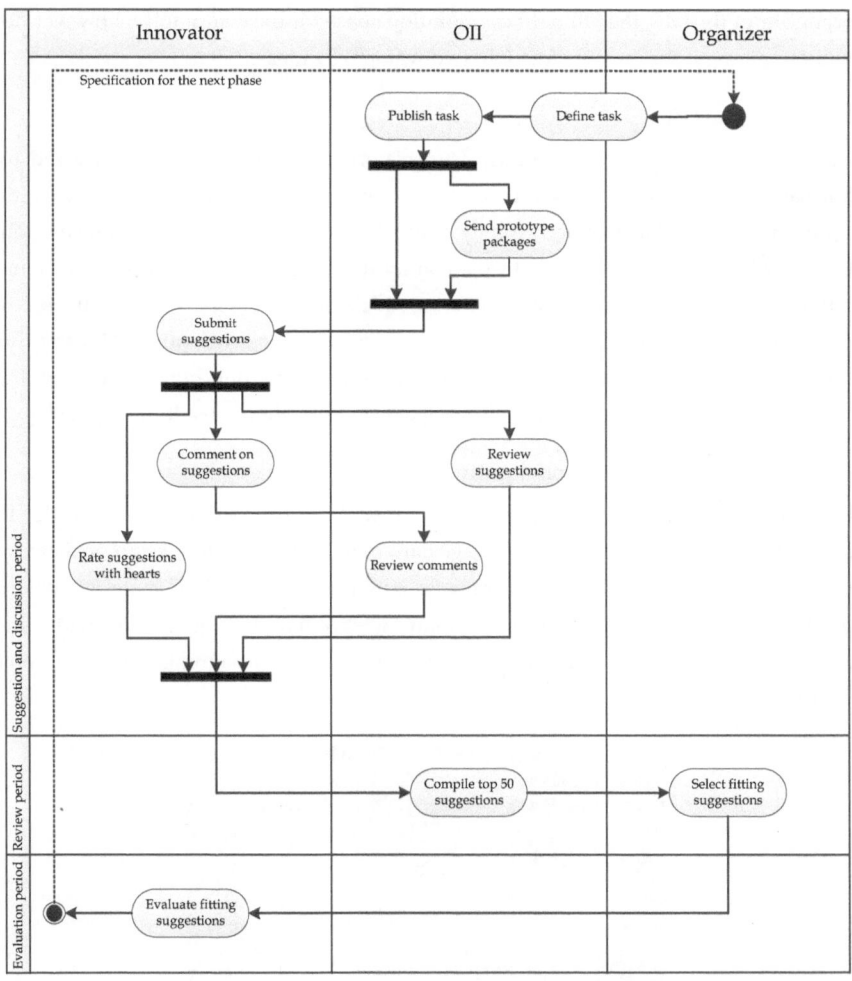

Figure 26: Activities of one phase in the unserAller development process

Suggestions and comments from the community

At the beginning of a phase, the community, i.e. the innovators, suggests solutions to the given task and comments on the suggestions. This way discussions arise, in the course of which suggestions are developed further (*suggestion and discussion period*). By assigning a precise task to a phase (e.g. "What scent does our shower gel have, does it shimmer and which color is it of?") in cooperation with its client, who is the organizer of the OIP, the OII defines a solution space for the community, however not without giving it the creative freedom to develop new ideas.

Review by the OII

All contributions by the community (suggestions and comments) are reviewed by innosabi. Redundant contributions are merged and enhancements of existing contributions are labeled accordingly and attached to the original contribution. Entries which do not comply with the community rules are removed (e.g. personal offences, unconstructive critique, contents which do not fit the topic or phase[524]). Users automatically receive e-mails about changes to their contribution. Beyond this review, innosabi answers administrative questions and questions regarding the content of a project. If the OII himself is not able to help, they consult the organizer.

Prototype packages for the community

Up to 750 prototype packages per project are sent out to the innovators in order to foster their creativity. Prototype packages are mainly used in the design or material phase. They contain basic components of a new product as well as materials whose purpose is to support the innovator's creativity[525]. While the organizer provides the content of the prototype packages, the OII takes care of designing, packaging and distributing them.

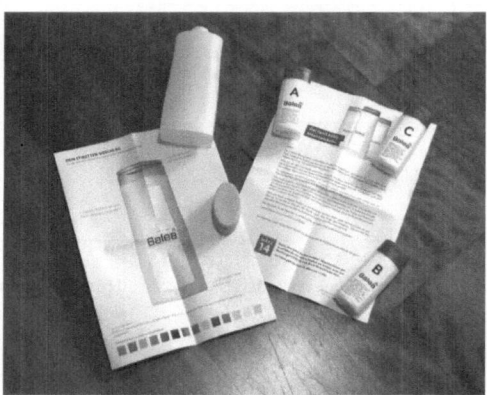

Figure 27: unserAller prototype package for the material phase of the Balea shower gel project

---

524 See https://unseraller.de/unseraller_rules; retrieved June 15, 2012.
525 See Figure 27.

The most active participants in a project can ask for these prototype packages in order to realize and test their suggestions. In order to receive the prototype packages, they have to run through an official application process. A prototype can be submitted as a suggestion using a configurator that offers for instance multiple options to configure a product (e.g. color, material etc.), photos and/ or textual descriptions. The available options to submit a prototype depends on the setup of the particular phase. By offering a configurator to the innovators, the technical feasibility of a suggestion can be granted (e.g. when a certain combination of fragrances is not feasible due to chemical constraints). On the one hand, the prototypes facilitate product development as well as depiction and evaluation regarding (interim) results. On the other hand, they are a means of motivation for the users:

> *In the meantime, we have a very standardized scheme how to put up a prototype package and we have a considerably detailed knowledge about which contents have which effects. [...] It has been always important that it looks good, that you enjoy participating. (I_CEO[526])*

Evaluation with hearts by the community

During a suggestion and discussion period, the community can rate suggestions with hearts in order to express their support[527].

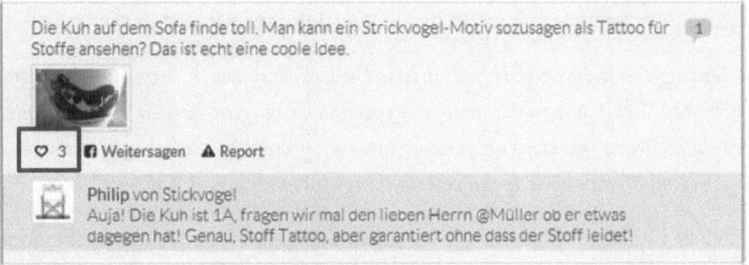

Figure 28: Evaluation of suggestions using hearts on unserAller

This evaluation has almost no effect on the sorting of the suggestions on the OIP. The entries are displayed primarily according to topicality and activity. The underlying

---

[526] This references the interview partner, as listed in section 1.3.
[527] See Figure 28.

activity index includes (but is not limited to) views, comments, and hearts, while the detailed algorithm is confidential.

> *In the first step, you receive completely free suggestions of the users who have got a pre-set solution space. During this phase, you can give little hearts to the entries. That is like Facebook's "I like"-button. You can give each entry one heart if you want to. This is no official evaluation, but rather an approval. Anyhow, we generate an internal ranking of the suggestions. (I_CIO)*

Review of suggestions by the organizer

After the suggestion and discussion period for the community, the OII presents the top 50 suggestions to the organizer. The selection is mainly based on the community's rating by hearts. The organizer reviews the suggestions and choses those that fit technically and strategically. Technical fit refers to producibility in the organizer's context:

> *[…] like for instance, a producer of yoghurt we talked to: he can process nuts on his machine, but he cannot process popcorn. No user can know this. (I_CEO)*

Apart from producibility, the new product has to fit in the organizer's portfolio strategically. All suggestions that are rejected by the organizer are substituted by new ones in order to hand back 50 suggestions to the community for the subsequent evaluation. Approximately, two to five suggestions are rejected in each phase.

> *During the suggestion period, it is not yet obvious which suggestions are among the top 50. It is possible that a suggestion is rejected or that it is just further behind. These two steps are a rather accepted process because everyone accepts that it is simply impossible to put certain things into practice. (I_CIO)*

This review by the organizer and its relevant stakeholders, facilitates the acceptance and producibility of the final product.

> *An unserAller project involves different departments which ex ante define their solution space and which also take the final decision in each phase. […] It is a relatively complex process. On the one hand, it has to be transparent and fair for the community, on the other hand, the client needs the opportunity to exert influence on the outcome if you get Pril with chicken flavor. (I_CIO)*

> *Due to the fact, that the community generates the ideas and the client provides the knowledge for production, the not-invented-here syndrome can be avoided. It is still*

the client's product because he himself advanced it that far. He still has to work on the product in order to make it producible. (I_CEO)

The OII deliberately establishes this division of roles between the organizer and the community. By this means, the community can freely generate suggestions and exploit their creativity without concerns about technical or strategic limitations. In the downstream review, the organizer adds its knowledge regarding technical and strategic feasibility.

Especially in the field of consumer goods, users will have creative ideas if you address a lot of them. Anyhow, the competence to produce a product is still our client's one. That is how we clearly separate producer and user competence. A producer of mustard knows best how to produce mustard. However, he does no know what his customers want. (I_CEO)

Evaluation by the community

After the organizer's review, the community evaluates the 50 remaining suggestions. Each innovator can allocate up to three stars to his favorite suggestions. The best suggestion - according to the community evaluation – wins and becomes an attribute of the new product and consequently a specification for the upcoming phases.

By dividing each phase into a *suggestions & discussion* and an *evaluation* period, innosabi sets the focus on the primary activity of each period and advantages for early suggestions diminish. Based on the introduction of the OIP, the following section addresses the OIP lifecycle management for it.

2.3 The OIP lifecycle management

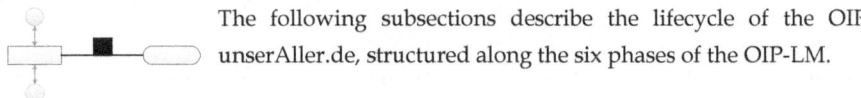 The following subsections describe the lifecycle of the OIP unserAller.de, structured along the six phases of the OIP-LM.

2.3.1 Requirements phase

The initial idea of unserAller originates from prior experience of the founders who have been consultants in open innovation for an average of more than five years. From their projects, they see a demand for user integration in the development of new consumer goods. They plan to build a platform to develop innovative products collaboratively that meet the consumers' needs.

Having this first concept in mind, innosabi intends to gather further requirements for their OIP. They therefore include both parties that are supposed to interact with the OIP primarily[528], namely potential organizers (i.e. producers of consumer goods, which are the OII's clients) and potential consumers of the organizers[529] (i.e. the innovators). At this point, it is an unstructured approach. From the first group of innosabi's potential organizers, they elicit requirements through short discussions with multiple companies and detailed discussions with single ones. They present their initial ideas and ask how they would design a process for the OIP and what they are concerned about. The organizers' major requirements include target group considerations and producibility of the results. From a target group perspective, the idea is to integrate all potential consumers, rather than only a particular group of them, e.g. designers or scientists, as this is the limitation in other innovation communities. The major aim is to get closer to consumers than traditional market research and innovation approaches could.

> *If you look at it from the perspective to get to know the user or that you want creative ideas, you will face a lot of frustration with traditional methods. If I do market research and statistics, I will have nice figures, but after all, it will not represent reality. (I_CEO)*

Additionally, innosabi finds that their potential organizers disapprove of using multiple tools to integrate consumers into their innovation process. They prefer a single tool for all consumer integration projects.

> *If you want to integrate users into an innovation process in a simple way, without the need to provide your own development team, and without using seven different tools and without having seven different user logins, that is what you get. You will be able to run every user-integration project with us. That is what it shall become. (I_CIO)*

Requirements elicited from the innovators' perspective represent a playful experience and the reassurance that their contributions are acknowledged and realized. In the end, they want to profit from the developed products. This leads to the necessity that products developed by the unserAller community are available in stores within a reasonable amount of time. This limits the selection of eligible innovation projects for unserAller which projects on convenience goods are particularly well suited for.

---

[528] See Figure 1 in Part I.3.
[529] Potential consumers of innosabi's clients are referenced as consumers from now on.

> *The basic idea of unserAller arose from the observation that people identified themselves with co-created products. We supposed that it would work even better for convenience goods. Meaning things, I build a relation to and with which I identify myself and my lifestyle somehow. Things that are quickly producible. Not like a car that needs 10-15 years of preparation but rather things we [in cooperation with the client] can bring to market within half a year. (I_CEO)*

Besides this basic requirement the consumers have, a proper graphical design as well as an easy-to-use and high-performance interface are requested. Owing to a close link with Facebook, security is also a concern and thus a requirement that has to be addressed.

> *Technology, design and the user interface are important factors to not lose a user. You cannot be good enough in them. Every user I bring to the platform costs money. If he does not get along or does not find functionalities he wants or the functionalities are not implemented well, I will lose him. (I_CIO)*

innosabi does not formally document requirements. They rather have a to-do list without detailed descriptions or dependencies. With these basic requirements and ideas, innosabi starts to design the first version of unserAller.

2.3.2   Design phase

 In the design phase, innosabi uses a perpetual-beta approach. They draw rough wireframes that show options to realize the requirements and focus on designing the social subsystem.

> *You can see it at this [screenshot of the] prototype. We have always been rather design driven. A good look was always important. You will have fun to participate and it will somehow work. [...] That is how unserAller was developed. A lot of prototypes and trial-and-error. Having a spark and then simply trying it. If it did not work, we leave it at that. That is it. (I_CEO)*

Discussions lead to a first design draft of unserAller. Right from the beginning, innosabi strives for a well-managed community and implements this idea into their design. Thus, from the beginning, they consider an administration backend to manage the community. They know from their experience, that this is crucial for an innovation community's success. The most critical design decision is the underlying innovation process. It takes long discussions and improvement cycles to design this aspect of unserAller with respect to social and technical feasibility. At the end, each process step of unserAller's development process matches one characteristic of the to-be-developed product. innosabi iteratively improves the development process until it meets the requirements. Its design is finalized prior to the initial launch of unserAller and is still valid.

> *We acted out the most important characteristics for 150, 200, unlimited products and estimated the decisions that will be taken in each phase. It always worked well. We said: 'Well, let us try it!'. It turned out that this process is usable for both, incremental changes but also disruptive new ideas. (I_CEO)*

Although the basic structure of the social subsystem is stable, details of it vary over time[530]: The first version of unserAller (Donkey) is a proof-of-concept for the development process, whereas the second version (Zebra) is a trial-and-error sandbox for details of the social subsystem around the development process.

Version	Donkey	Zebra	Pegasus
Release date	June 2010	February 2011	October 2011
Architecture	Plain PHP, Plain MySQL	Plain PHP, Plain MySQL	ZEND-framework, MVC, ORM, Magento
Purpose	Proof of concept with basic functionality	Graphical redesign and enriched functionality	Technical redesign; secession from Facebook; my unserAller
Origin of version name	Rather gray graphical design (like a donkey)	Graphical design dominated by dash lines (like a zebra)	Slogan: "Because we do not need a book to fly" (I_CEO)

Figure 29: Major versions of unserAller

During the runtime of the first innovation project, innosabi had the idea to distribute prototype packages to connect the online and the offline world. Today, this is one of the key value propositions of unserAller.

> *What motivates them is fun doing it. If you get such a package and can try out things, it becomes very playful. There is this term of gamification and that is a valid point. If you enjoy doing it, you will do a lot. (I_CEO)*

In addition, with the experience from the first projects, the administration backend is refined to meet the requirements of the community managers (i.e. innosabi's

---

[530] See Figure 29.

employees) in order to facilitate their work. The third version of unserAller (Pegasus) keeps up the social subsystem of Zebra and adds scalability – as a major new requirement from a business perspective – by a software-as-a-service (SaaS) version of unserAller. With this SaaS version, organizers can set up and manage innovation projects in the unserAller front-end without help from innosabi. This case description applies the perspective of the Pegasus version (and if not differently stated) without its capability to run multiple concurrent projects[531].

The respective technical subsystems are designed to complement the purpose of the different versions. This means that Donkey and Zebra are designed in a very lean way with plain PHP[532] and MySQL[533]. This enables flexibility and quick responses to a volatile social subsystem.

> *It was an extremely quick development and we did not know what the outcome will be. We started from scratch for our [project name] and [project name] project. We did not know if and how such a project will work. That is why we had to make a great deal of quick changes to the system. That called for plain PHP programming. (I_CIO)*

This is also the reason, why the first two versions of unserAller are integrated with Facebook: This allows quick and easy access to user authentication and a rapid community building. Once the social subsystem stabilizes in Zebra, the architecture is adapted accordingly: Pegasus is redesigned from scratch to complement the now stable social and technical requirements while building upon state-of-the-art technical concepts.

> *We now know what the product [unserAller] will look like and have the vision how it shall be and will be. We did not have that for the first version. [...] At the beginning of the year, when large clients approached us, scalability became a subject [...] and that is why a complete redesign from scratch was justifiable anyway. (I_CTO)*

---

[531] This focus allows to clearly outline the role of the OII.
[532] PHP is an open source scripting language to generate dynamic websites; www.php.net; retrieved June 15, 2012.
[533] MySQL is an open source database management system; www.mysql.com; retrieved June 15, 2012.

innosabi draws on an object-oriented programming paradigm using ZEND as a framework[534] with a model view controller (MVC) architecture[535] in PHP, Doctrine2 as an object relational mapper (ORM)[536] to design the object-oriented database structure and jQuery[537] for the user interface. Additionally, Magento[538] is selected as a shop system because it is based on the ZEND framework as well. Website tracking is realized by Piwik[539]. Also the close link to Facebook is dissolved and replaced by a proprietary user management using the ZEND framework. By this redesign, innosabi aims at harmonizing all technical components.

2.3.3   Build phase

Three developers of innosabi work full time on the development of unserAller. Outsourcing of the development is not an option, as innosabi considers providing technology as one of their USPs. They consequently want to keep their core competency in-house.

> *We actually have a high technological competency in our team. We can quickly realize many things by ourselves and iterate them and try out things and see what works best. That is possible because we do not always have to ask an agency to program it for us. [...] That was actually key to bring unserAller to this technical realization that it is today. (I_CEO)*

> *I cannot exactly tell how long [the designer] needs for a design. But I know that none of the designs that he delivers will be the same in the next week. Once it is live, we act very fast. And that is naturally a benefit. We are a young company. We have to be able to act fast. (I_CTO)*

Although they do not explicitly follow a particular method, they coordinate their work using a Scrum-like technique with Post-Its[540]. Tasks are sorted by priority and

---

[534] ZEND is an object-oriented framework for websites using PHP; framework.zend.com; retrieved June 15, 2012.

[535] MVC is a design pattern for applications to separate the model (program logic), view (output to the user) and controller (input by the user). See Fowler (2002).

[536] Doctrine2's ORM is used to design a database's object-oriented architecture according to applicable standards; www.doctrine-project.org; retrieved June 15, 2012.

[537] jQuery is a JavaScript library to design cross-browser compatible user interfaces; www.jquery.com; retrieved June 15, 2012.

[538] Magento is an open source e-commerce system; www.magentocommerce.com; retrieved June 15, 2012.

[539] Piwik is an open source web analytics software; www.piwik.org; retrieved June 15, 2012.

[540] See Figure 30.

scheduled to a particular time frame. If someone takes care of a task, they put the Post-It on their computer screen to show their current tasks to everybody.

Figure 30: Scrum-like coordination of tasks at innosabi

Tasks are mainly assigned by competence: one developer takes care of the user interface while the other one deals with the database and program logic. The third developer focuses on bug fixing. The codebase is stored in a repository (i.e. Beanstalk[541]) that is linked to a development and live system of unserAller. Critical functionality is developed using a 4-eyes principle. Ideas are implemented as rough prototypes prior to the final implementation. This allows innosabi's employees and beta testers to test, evaluate and optimize new functionality prior to its final implementation.

All functionality of unserAller is implemented as modules. Thereby, usage of existing components and reusability are key concepts. *Usage of existing components* is for instance reflected by external open source software that is integrated into unserAller: In order to benefit from best practices, innosabi integrates existing tools like the shop system Magento or tracking system Piwik. After the initial work to integrate these external components, this allows an easy implementation of additional functionality that is not yet in the scope but available in the external tool (for instance a new tracking measure). *Reusability* is facilitated by the modular architecture of unserAller. Projects are for instance listed on the homepage, on the project overview and on an organizer's profile. All these different views on projects are based on the same module.

---

[541] Beanstalk is a subversion system to collaboratively work on source code and deploy it; beanstalkapp.com; retrieved June 15, 2012.

Quality is assured by intensive testing especially with respect to cross browser compatibility. In a first step, innosabi's developers mutually test new functionality. In a second step, non-technical employees like the community mangers perform tests.

> *We mostly identify bigger bugs internally if we give it [the new functionality] to [a non-technical employee]. He just needs five clicks to find something we have never seen before. He is our best bug reporter. (I_CTO)*

If these internal tests are successful, innosabi draws on external beta testers.

> *We have a list of 15 beta testers. Those are mainly companies, which are completely different. We have software developers who want to test the interfaces, caterers, small producers, large producers and also community members. (I_CEO)*

Bugs are often solved within minutes. This also applies to small feature requests. Bigger requests are prioritized within the team. While frontend changes are prioritized, the development of the administration backend is not neglected.

> *If we develop new functionality, at the same time we will create a corresponding backend interface. Otherwise, we would not be able to handle the user support if they do not find something. [...] [From an IT perspective] each calling user distracts our lean processes. (I_CIO)*

2.3.4   Deploy phase

 Due to the managed code repository, the technical deployment of unserAller can be done from Beanstalk. This is also required, as smaller changes are published immediately in a perpetual beta paradigm.

> *There are always those little bugs and changes, [...] but we can usually solve and deploy them within minutes. (I_CTO)*

Next to these small updates, bigger updates (like for instance the introduction of a new online shop) are deployed according to a milestone plan. Only major updates are communicated in newsletters. Smaller changes occur in a silent rollout in order to reduce spam to the community.

The initial deployment of unserAller took place at a launch party with stakeholders. Prior to the launch, the first innovation project is set up. During the launch party, the innovation project is occupied with content in order to have initial content when the online community starts to work on it. Marketing for unserAller is carried out via Facebook. innosabi uses the organizer's Facebook communities to address potential innovators. This way, they nurture their own innovation community without spending significant resources:

> *We were able to build the community really fast without spending a single Euro for advertising. This was actually supported by the viral effect on Facebook. (I_CEO)*

Key to this marketing concept is access to organizers with Facebook communities which innosabi tackles actively. Besides the rollout to the community, innosabi takes care of a proper integration of the community's results into the organizer's organization. Activities include trainings on how to use and what to expect from unserAller, newsletters to employees and videos messages to promote the innovation project in an organizer's intranet. This is done to facilitate acceptance and later on realization of the products developed by the community.

2.3.5 Operate phase

A proper management of the unserAller community is a core value proposition of innosabi. In order to run innovation projects within the community, innosabi employs dedicated community managers that ensure ten hours per day, seven days a week community management. Each contribution by the community is reviewed and modified or deleted if necessary[542], as the following example shows:

> *We had this suggestion in our snack project. A snack for 'after having sex'. It was of course very popular and many users clicked on it because they thought it is cool. However, this is no need of the consumers. You can see it from the comments. That is why we deleted it. [...] and this is accepted by the community. (I_CEO)*

The review is supported by innosabi's administration backend. The administration backend for instance indicates all open review tasks, identifies duplicate suggestions to propose a merge of them and allows to create new innovation projects or to modify current ones[543].

---

[542] See subsection 2.2.2.
[543] See Figure 31.

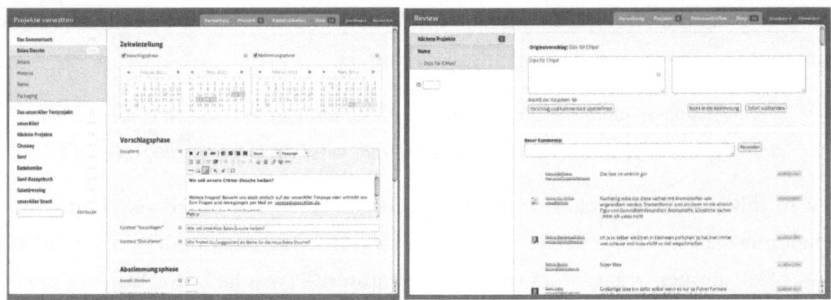

Figure 31: Administration backend of unserAller (version Zebra)

The administration backend is required in order to deal with up to 2,300 suggestions and 6,600 comments per innovation project. Beyond this review, innosabi sends newsletters to active participants of an innovation project, answers administrative questions and inquiries regarding the content of an innovation project. If innosabi is not able to help, they consult the organizer. All of this is intended to keep up a close link to the community. This also includes proactive personal approaches towards selected community members.

> *We read every contribution of the community, review it and answer every single question. If we have 450 design proposals and 2,500 comments like in the [name of project] project, we have to approve all of them. This is quite a big effort, but that guarantees us that we can quickly react on questions, that we can avoid disputes and that the community feels taken seriously. [...] but it also deals with remaining a feeling for the community. Thereby, we can realize if something gets out of control. (I_CEO)*

Besides the community management, innosabi takes care of frequent reports on each phase's activities and a proper preparation of the community's overall result for the transfer to the organizer.

2.3.6   Optimize phase

 As already outlined in the design phase, innosabi continuously optimizes unserAller. This does not only apply to bugs but also to new features. Thereby, suggestions for optimization do not only origin from innosabi's employees, but also from externals. External requests origin from both, their community and organizers. Although a standardized feedback functionality exists, most feedback from externals is given personally. This is a result of the close link to the community.

To develop new functionality, innosabi tests it with a single organizer or user and transfers the outcome to others.

> *[The client] tells us which functionalities he wants and how he imagines the project. Then we build it the way that it is perfect for him and hope that it is transferrable to our other clients. (I_CEO)*

innosabi runs a perpetual-beta approach with multiple iterations to develop new functionality. They are especially striving to integrate functions that can enhance their business model or foster community activity. Anyhow, they try to retain basic functionality in order to avoid confusion:

> *We try to not break functionality, which is of course not always possible. […] We try to silently introduce new functionalities. Now and then, there is a new button and you can try it. (I_CIO)*

Besides feedback from externals, innosabi also uses tracking analyses to identify problems and evaluate new features.

> *We do not always activate the heat map [of Piwik]. We only activate it if we integrate something new in order to evaluate it. (I_CTO)*

To sum it up, unserAller evolves in small steps and some big leaps of which the latter ones are marked by milestones.

> *From a development perspective, it [unserAller] evolved step by step with the projects we ran. We had a vision of a platform in mind right from the outset. Nevertheless, we said that the vision is too big for a first step and that we need to gain experience first. (I_CEO)*

2.4 Summary and conclusion

The following subsections summarize findings and draw conclusions from the innosabi case concerning the three facets of OIP lifecycle management, namely the OII (*subsection 2.4.1*), the OIP (*subsection 2.4.2*) and the OIP lifecycle management (*subsection 2.4.3*).

2.4.1 The open innovation intermediary

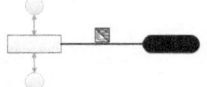 innosabi is an OII that focuses on producers of fast moving consumer goods. They use their standardized OIP to integrate consumers in the development process for new products.

The OII aims at transferring tacit knowledge from the innovators, to the organizer. As most of innosabi's clients focus on consumer goods, the OII helps organizers to articulate demands of consumers. By providing a standardized community, the OII allows an organizer to draw on diverse knowledge to solve an organizer's innovation problems.

The constant review of all contributions on the OIP ensures an integration of similar ideas that stem from the community in order to make them valuable for the organizer. The offering of products that were created by the community fulfills a twofold role. On the one hand, the community members, i.e. the innovators, can see the result of their innovation activities, be proud of it and use it if they wish to. Thus, it fosters motivation. On the other hand, offering this additional distribution channel helps organizers to sell their new product.

2.4.2   The open innovation platform

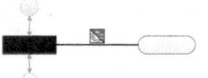 innosabi created a community-based OIP to develop new products collaboratively applying a standardized development process. All innovation projects of innosabi's clients run within the default graphical design of the OIP. Although there are endeavors to add new channels to run external innovation projects using unserAller (e.g. by embedding buttons on an organizer's website), the hosted solution is still the major case in use.

The design of the OIP is driven by the aim to playfully generate innovative products that are accepted and producible by the organizer of an innovation project (i.e. innosabi's client) while the community should maintain a sense of ownership of the results. This is realized by a development process of unserAller which accomplishes two major aims: Firstly, the designed development process integrates all three steps of the innovation process including the search, selection and implementation, while all of them are supported by a community that addresses a broad audience (including for instance interested amateurs[544]). In order to integrate the community in the implementation phase of the innovation process, innosabi developed prototype packages which are mailed to the innovators. Thus, compared to traditional OIP projects that allow for written contributions only rather than (physical) prototype generation, more tacit knowledge is transferred to the organizer. With this development process innosabi created a seminal way to integrate the virtual

---

[544] Hallerstede et al. (2010).

(online) and physical (offline) world. In addition to the functions of prototypes as they are described in the literature[545], innosabi uses the creation of prototypes to motivate innovators. This approach represents a community-based, decentral and inexpensive alternative to the lead user method[546].

Secondly, in each step of innosabi's development process, evaluation mechanisms are used in order to not only generate contributions, but also to have a community-based selection of them and a corresponding transparency. Applying an interim evaluation by the organizer, they retain an opportunity to influence the outcome of an innovation project. This task division gives the community creative freedom. At the same time it sets up guidelines that guarantee the strategic and technical fit of the developed product to the organizer and reduces the risk of the not-invented-here syndrome.

The task division between organizer, community and OII[547] is of particular interest. As innosabi takes care of the innovation project design, the interaction with the community and the processing of the results to the organizer, there is no need for the organizer to hold or build up knowledge concerning open innovation (e.g. expertise on the management of an innovation community). The OII instructs the organizer according to all tasks it has to accomplish.

2.4.3 The OIP lifecycle management

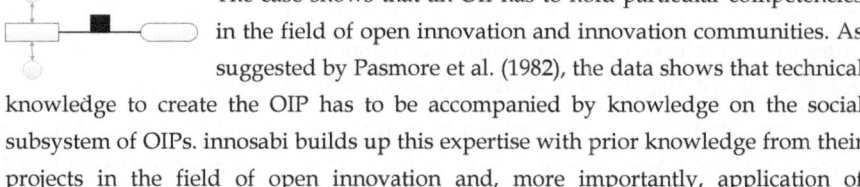
The case shows that an OII has to hold particular competencies in the field of open innovation and innovation communities. As suggested by Pasmore et al. (1982), the data shows that technical knowledge to create the OIP has to be accompanied by knowledge on the social subsystem of OIPs. innosabi builds up this expertise with prior knowledge from their projects in the field of open innovation and, more importantly, application of continuous improvements. They additionally consult experts in the field of open innovation to learn from latest research and practice.

The technical subsystem builds the foundation for innosabi's success. By applying state-of-the-art technology and latest software design concepts, they realize a modular and flexible architecture of the OIP that enables innosabi to react quickly to new requirements without compromising the core architecture. The basic principles

---

[545] See Bullinger et al. (2011); von Hippel and Katz (2002).
[546] For details on the lead user method see Luethje and Herstatt (2004).
[547] See Figure 26 in section 2.2.2.

innosabi consequently applies are modularity, flexibility, usability (dominated by simplicity and intuitive navigation), reuse of existing external components like frameworks, and application of state-of-the-art programming paradigms. On the one hand, this leads to a raised effort to assess the best approach to implement a requirement, e.g. due to required desk research to identify and evaluate options, but on the other hand this reduces required maintenance and grants flexibility to react to a changing social subsystem.

Additionally, all functionality the OII implements is accompanied by a corresponding administration backend. This enables scalability, as innovation projects can be set up, configured and maintained within minutes. The arrangement, runtimes and configuration of the single phases are flexible. Consequently, an innovation project can be customized according to an organizer's requirements, while the benefits of the standardized design of the OIP remains.

innosabi strives for a superior user experience in order to foster voluntary contributions of the innovators. Consequently, they try to reduce distractions like bugs to a minimum by applying quality management. That is why all new functionality is evaluated in a three-step process that includes testing by peer developers, community mangers and external beta testers.

While striving for a state-of-the-art technical subsystem as an enabler that reduces barriers of participation and enables scalability, innosabi focusses on community management which they see as the crucial activity in an OIP's lifecycle. In order to accomplish community management, innosabi's core asset is a specialized administration backend that supports the community management tasks.

This administration backend is also used to carry out all other administrative tasks to run unserAller, like the creation of new innovation projects or traditional content management. This allows for instance for a quick and low cost setup of new innovation projects which finally facilitates scalability. This technical basis of innosabi paired with a standardized development process, which can be configured and elaborated in a single workshop with the organizer, allows for a short lead time to start an innovation project on unserAller. This is particularly interesting, as the effort for an organizer to run a custom innovation project is reduced to a minimum. Due to the fixed schedule of each innovation project on unserAller, using unserAller is a very structured and projectable process for the organizer, which is additionally strongly guided by the OII.

The social subsystem of unserAller is designed in an iterative process. innosabi heavily integrates external ideas to improve it. Therefore, they gather feedback and ideas from experts, organizers and the community. Promising feature requests are evaluated in pilot innovation projects in order to learn from them. If they work as expected, they are rolled out to all innovation projects. Thus, details of the social subsystem of unserAller are in constant change while the basic structure remains.

Whatever innosabi wants to implement in the social subsystem has to be mastered from a technical point of view. This is allowed by a high technical competency in the team paired with the freedom to do a proper, rather than a quick implementation of new functionality.

The core value proposition of innosabi is its competence to manage their innovation community. Based on their experience and iterative improvements, they developed a successful way to manage their community. Due to the appropriate technical basis, they can spend their resources in community management and other core tasks, rather than providing technical support or dealing with bugs.

According to the described social focus in the OIP lifecycle management activities, innosabi focuses on the following phases of the OIP-LM framework: A lot of effort is put in the assessment, design, evaluation and management of the social subsystem of their OIP in order to foster both, the voluntary contributions of the community and a viral marketing effect. Accordingly, after the initial introduction of unserAller, the phases *operate* and *optimize* are emphasized by innosabi. Table 19 summarizes the key findings of the innosabi case.

Table 19: Key findings of the innosabi case

	OIP-LM focus	Operate and optimize
	Key findings	Focus on the design and management of the social subsystem
		Most resources invested in community management
		State-of-the-art technical subsystem
		Sophisticated administration backend to set up innovation projects and to support community management
		Striving for and reaching scalability
		Self-perception: technology provider that adds open innovation consultancy
		Viral marketing via organizers' Facebook communities

3   Case 2: HYVE

The second case investigates the open innovation intermediary HYVE with its OIP HYVE IdeaNet©. Table 20 summarizes the key facts on the HYVE case. The structure of this case follows the structure described in chapter 2.

Table 20: Key facts on the HYVE case

OII	HYVE AG (referenced as HYVE)
Description	innovation agency with multiple approaches to open innovation including HYVE IdeaNet© to design custom innovation contests an innovation communities
Foundation	2000
Employees	> 40 (including 2 founders)
Homepage	www.hyve.de; www.innovation-community.de
OIP	HYVE IdeaNet© (instantiation for instance www.gemeinsamselten.de)
Tools	innovation contest; innovation community

3.1   The open innovation intermediary

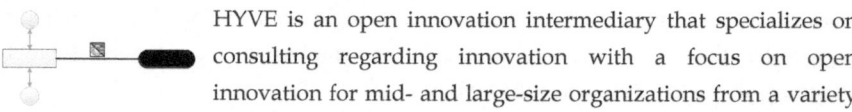

HYVE is an open innovation intermediary that specializes on consulting regarding innovation with a focus on open innovation for mid- and large-size organizations from a variety of industries in Germany. The OII, which was founded in 2000, is one of the first movers, most established and market leading consultancies for open innovation in Germany[548]. It is spilt into three business units which also mark HYVE's core competencies: The *HYVE Innovation Community* (HIC) focusses on the creation and management of OIPs to integrate innovators into an organizer's innovation process. The *HYVE Innovation Research* (HIR) team specializes in techniques that systematically search for innovation in secondary data, for instance by doing Netnography in user communities[549]. The *HYVE Innovation Design* (HID) team focusses on industrial and web-design as well as lead user and ideation workshops. The present analysis focusses on a *HIC perspective* although the three business units

---

[548] Williams (2011).
[549] See Bartl and Ivanovic (2010).

closely work together in order to realize synergies if an OIP project requires this. In order to remain state-of-the-art, HYVE's strategy includes a close link to the open innovation research community. They both absorb latest results of research and contribute to them. Therefore, the OII collaborates with several research institutions, develops new open innovation methods (like Netnography[550]), visits conferences and write research papers[551].

Regular clients of HIC include companies like BMW (example OIP project: BMW Group Idea Contest Tomorrow's Urban Mobility Services[552]), Siemens (OSRAM LED - Emotionalize your light [553]) or Swarovski (Enlightened Jewelry Design Competition[554]), from whom each OIP is designed and realized individually. A list of reference OIPs can be found on www.innovation-community.de. HIC employs a young and multi-disciplinary team of creative people who conduct custom projects in the field of open innovation from the initial idea until the transfer of the results to the organizer. Today's focus of HIC lies on OIP projects to integrate employees or consumers into an organizer's innovation process. The framework they use to design OIPs, namely the HYVE IdeaNet©, is introduced in the following.

3.2   The open innovation platform

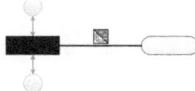 The following two subsections elaborate on the HYVE IdeaNet© (*subsection 3.2.1*) as well as its processes and core functionalities (*subsection 3.2.2*).

3.2.1   Overview

The HYVE IdeaNet© is a framework that can be used as a basis for an OIP. This framework contains a set of basic functionality (e.g. user registration, submission of ideas and evaluation mechanisms) that is used in all of HYVE's OIP projects. In addition, the framework has the opportunity to use advanced existing functionalities or to build new ones (e.g. configurators, serious games). The layout and graphical design as well as the detailed architecture are customized according to an organizer's needs. Each OIP that is realized using HYVE IdeaNet© runs on a distinct domain and webserver. The HYVE IdeaNet© can be used for both (organization-)internal and external innovation communities as well as for innovation communities that serve

---

[550] See Jawecki and Fueller (2008).
[551] For a list of research papers see www.hyve.de/publications.php; retrieved May 22, 2012.
[552] www.bmwgroup-ideacontest.com; retrieved May 22, 2012.
[553] www.led-emotionalize.com; retrieved May 22, 2012.
[554] www.enlightened-jewellery-design-competition.com; retrieved May 22, 2012.

both purposes. HYVE usually integrates traits of innovation contests in their innovation communities. The OIP Gemeinsam fuer die Seltenen[555] (GfdS) is a typical representative of a HYVE IdeaNet© implementation. It is used as an example OIP project to illustrate HIC's approach to the design and management of OIPs.

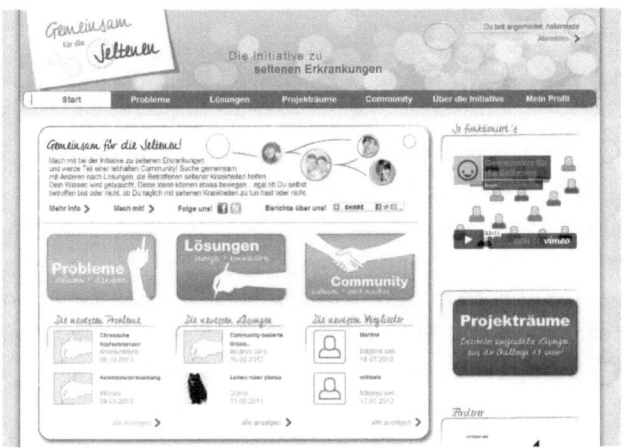

Figure 32: Homepage of GfdS's HYVE IdeaNet© implementation

3.2.2   Processes and functionality

Gemeinsam fuer die Seltenen is an innovation community based on the HYVE IdeaNet© which was launched in March 2011. Its purpose is to identify problems of individuals with rare diseases in order to collaboratively develop product or service innovations to solve these problems. More than 1.300 patients, caregivers, health workers, physicians, nurses (i.e. core inside innovators); family members, friends, fellow patients (i.e. peripheral inside innovators); or researchers, engineers, product managers and civil servants (i.e. outside innovators) are participating[556]. Innovators can provide an extended profile to share their medical history or expertise, interact with other community members and discuss general topics in forums. Additionally, users can post their current state of health (like "I'm feeling very well today" or "I have problems with my respiration today") to share it with the community. Other community members can see an individual's status and the community's overall status. Figure 33 summarizes the activities in the GfdS community.

---

[555] www.gemeinsamselten.de; retrieved May 22, 2012. See Figure 32.
[556] For the definition of the types of innovators see Neyer et al. (2009) and Part II.1.2.

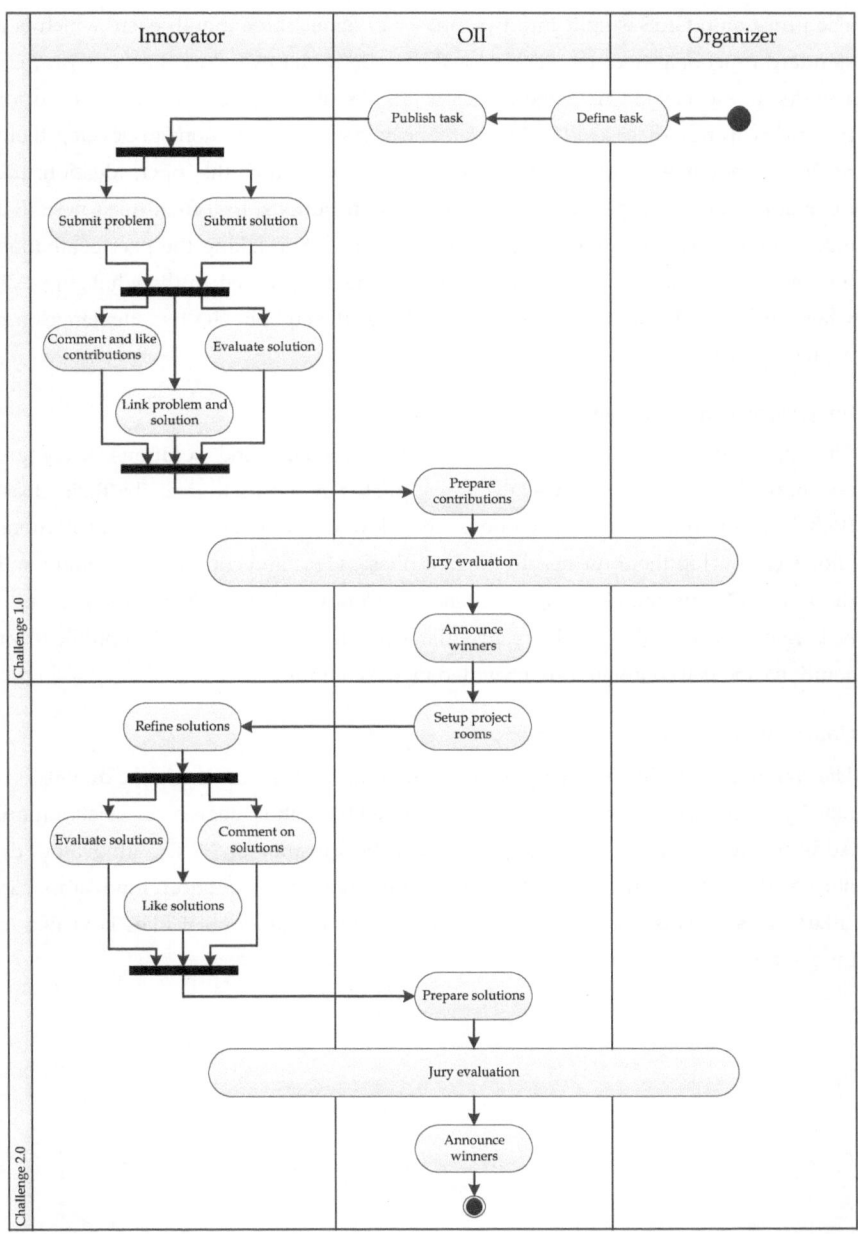

Figure 33: Activities in the GfdS community based on the HYVE IdeaNet©

The process on GfdS is split into two phases of about three month each, which is a standard functionality of the HYVE IdeaNet©. Both phases follow the same process. The first phase, called Challenge 1.0, serves to identify problems of and solutions for patients with rare diseases. The innovators can discuss submissions to develop them further. If possible, solutions are linked to problems. After this open ideation, the innovators perform a peer-evaluation of the contributions. In a final third period, a jury of experts selects three winning contributions while taking the peer-evaluation into account. Winning contributions of Challenge 1.0 proceed to the Challenge 2.0, where they are elaborated further. The activities throughout the two challenges are explained in detail below.

Suggestions and comments by the innovators

The innovators can provide suggestions for problems and solutions. They are encouraged to provide text, visualizations, photos or videos to describe their ideas. Both team or individual contributions are allowed. Commenting on contributions allows to develop them further. In the first phase, ideas and concepts are sought and discussed. During the Challenge 2.0 concepts from Challenge 1.0 are advanced into elaborated concepts that can be realized later on. In total, more than 200 problems or solutions and 900 comments are provided by the community.

Community evaluation

The community evaluates suggestions on a five star scale according to the dimensions novelty, usefulness, feasibility, market potential and degree of elaboration. Additionally, an overall rating is given for a short evaluation[557]. The innovators do not see the evaluation results. Apart from this five star evaluation, innovators can "like" ideas. This is not an official evaluation, but the amount of likes is visible to innovators.

---

[557] See Figure 34.

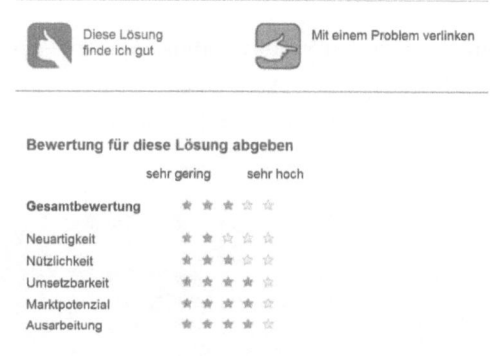

Figure 34: Evaluation of submission at GfdS based on the HYVE IdeaNet©

Jury evaluation

Subsequent to the community contributions and evaluations, a jury of experts from the field (including patients, physicians, open innovation experts, etc.) selects winning contributions through the application of an online evaluation. The jury considers the peer-evaluation, but maintains the freedom to select the most promising contributions.

Activity index

Innovators receive activity points for each activity on the OIP. Activities include for instance posting a solution, commenting and evaluating. As shown in Table 21, each activity has its particular amount of points that is awarded to an innovator. Activity points are viewable on a innovator's profile. Additionally, the most active innovators are highlighted on the community homepage.

Table 21: Points awarded for activities on GfdS

Activity	Points awarded
Submitting a problem	10
Submitting a solution	20
Commenting on a problem or solution	2
Writing a personal message	1
Posting today's state of health	1
Short evaluation of a submission	1
Detailed evaluation of a submission	3

User workshops

Aside from the online advancement of ideas, offline user workshops are conducted to elaborate ideas in the Challenge 2.0[558]. Workshop participants are recruited from the most active contributors to an idea. Results of the workshop are channeled back to the online community.

Figure 35: User workshops for Challenge 2.0 of GfdS

Awards

The innovators of the five best ideas according to the jury evaluation receive monetary awards (including iPads, iPods, digital cameras, backpacks, …) after each of the two challenges. Additionally, the three most active innovators are rewarded with prizes. The innovator who was ranked best may select a prize from a pool of prizes followed by the second best and so on. At the end, the most active innovators may select from the remaining prizes, again in order of their ranking. Now that the OIP has been introduced, its lifecycle management is addressed in the following section.

---

[558] See Figure 35.

3.3   The OIP lifecycle management

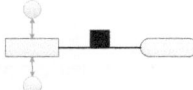 The following subsections describe the lifecycle of an OIP that is based on the HYVE IdeaNet©. This section is structured along the six phases of the OIP-LM.

3.3.1   Requirements phase

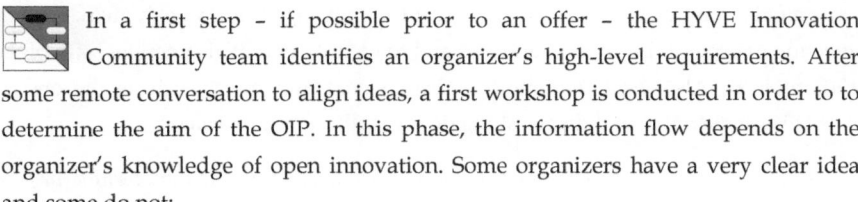 In a first step – if possible prior to an offer – the HYVE Innovation Community team identifies an organizer's high-level requirements. After some remote conversation to align ideas, a first workshop is conducted in order to to determine the aim of the OIP. In this phase, the information flow depends on the organizer's knowledge of open innovation. Some organizers have a very clear idea and some do not:

> *We have two types of clients. The one type approaches us and wants us do build an innovation community platform. This is a typical me-too, as others have it as well. [...] The other type of client approaches us and says: this is our process and this is what the tool should look like. They have some 50 pages requirements document because they invested months in thinking it through. (H_CPM)*

In the case of a well-informed organizer, the OII's major task is to figure out the core aim of the OIP and to prioritize requirements in order to cut non-crucial ones. In the second case, the organizer is consulted regarding an appropriate use of OIPs.

Besides requirements regarding the user interface, requirements regarding the administration backend are elicited. Each organizer has its interest in some particular details. Thus, corresponding reports have to be considered. Eventually, the OII has a list of high level requirements including "What is the aim of the OIP?", "Who is it for?" and "What kind of ideas and inputs are sought?". Requirements elicitation focuses on the social subsystem, rather than the technical one.

3.3.2   Design phase

 Based on the requirements elicited in the preceding phase, the project manager at HIC works on a concept for the OIP. For this, his experience in the field is crucial.

> *One situation is really dangerous: those [clients] who believe to understand the topic and then try to process it in their traditional innovation process. For instance seeing an innovation contest as an idea production machine. They think that you*

get three new ideas at the end of the day that they can put into their traditional innovation process. (H_SPM1)

This experience is based on an average of four years in open innovation consultancy and at least eight years of experience for HIC's senior project managers. In any case, due to the complexity and amount of options in designing an OIP, HIC does not rely on a single point of view. Each new concept is challenged in a weekly management meeting in order to proof its soundness and to identify potential synergies with other OIP projects. Also legal requirements are covered and traps, such as the one in Henkel's chicken flavored dishwashing detergent[559], can be revealed:

For instance Pril. What we [would have done] differently, is to implement a jury evaluation into the platform's process in order to neutralize the communities' subjectivity. That would not have happened to us because we understand how communities function, why they do what they do and how they are motivated. (H_SPM1)

With a concept in mind, a second workshop with the organizer is performed. The major task of this workshop is the definition of the concept and the functions of the OIP. Therefore, more details like the exact topic, rules, roles, rights and processes are discussed.

Figure 36: Design workshop at HYVE using metaplan technique

---

[559] See Spiegel Online (2011).

The workshops are moderated using metaplan technique and pre-labeled areas (e.g. for functions, roles, rules) on the wall. On the one hand, this course of action gives the workshop some structure with topics that have to be covered, on the other hand, it preserve room for creativity[560]. The major task HIC's project managers have is to maintain a simple concept that fosters open innovation mechanisms:

> *The challenge is to keep it as simple as possible. That is a key learning from the [project name] project: simpler. Work on the logic that enables the viral mechanisms and then start it. (H_SPM2)*

> *In general we try to have a platform that is as lean as possible. That means: less is more. If our client says that is a nice module, we would like to have that, then we say: leave it, we should only include things we really need. That is our unique selling proposition. (H_CPM)*

To ensure the OIP's success, HIC's project managers state and defend their point of view in a clear way if it comes down to disagreement with the organizer.

> *The client gathered, if at all, experience from a single platform. That is why we say sorry, that is not reasonable and it will not work like this. (H_CPM)*

At this point, HIC distinguishes between two types of OIPs: internal and external OIPs. Internal ones are for an organizer's employees only (core inside and peripheral innovators) and external ones are those to integrate innovators like customers (outside innovators)[561]. Depending on the type of OIP, different actions have to be taken. The focus for internal OIPs is set on the identification of stakeholders, corporate design guidelines, hosting, integration into the organizer's corporate landscape and technical question. The focus with external OIPs, on the other hand, lies on graphical design, marketing and target group considerations. Additionally, although the underlying principles are the same, they have to be addressed differently and reflected by the concept of the OIP.

> *On the one hand, external ones [OIPs] are completely voluntary. It is for fun only. Internal platforms on the other hand, inherit some kind of pressure from the managers or executives that force employees to use it. Those are two completely*

---

[560] See Figure 36.
[561] For the definition of the types of innovators see Neyer et al. (2009).

different [motivational] approaches. [...] The levers are the same but you have different mechanisms. (H_PM2)

Although the single functions are easy to implement, their interplay has to be designed carefully, as the incentives for participation illustrate.

Then there is the works council. They say, if employees submit cool ideas, they have to get incentives. We said that that is not possible, as we want to foster collaboration. If you provide incentives, each employee works on his own credit and no one writes comments on other's ideas. That is why we said we go without. [...] We finally got there and are also backed by the works council by now. (H_CPM)

When the project manager has the rough concept at hand and figured it out in mind, they create wireframes for another round of alignment with the organizer[562]. Wireframes in HYVE's conception depict the OIP's layout in order to show the arrangement of the content as well as the interface elements and their interplay. Once the wireframes are approved, the next process step to build the OIP is initialized.

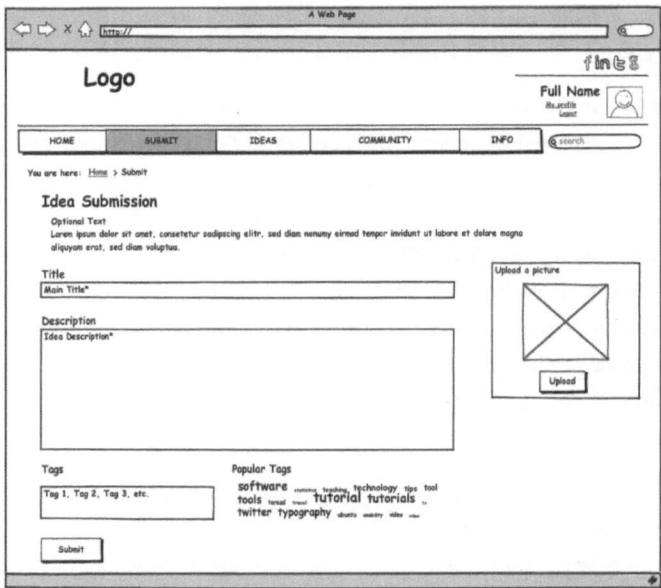

Figure 37: Wireframe of an internal OIP using HYVE IdeaNet©

---

[562] See Figure 37.

3.3.3 Build phase

 The first step in the build phase is to create mockups from the wireframes. Mockups are design prototypes that serve to evaluate the graphical design. Therefore, the project manager works together closely with a web-designer. The web-designer also has a background in open innovation, which facilitates communication between designer and project manager. Additionally, the web-designer stays in close contact with the HID department in order to align state-of-the-art design principles. The mockups have to be accepted by the project manager in a first step and by the organizer in a second one. This way, guidelines that belong to the organizer's corporate identity are considered in particular in order to grant identification with the organizer's brand. If adjustments are required from any party, the OIP-LM process might revert to the requirements or design phase.

Based on the mockups and wireframes, which specify the graphical design respectively the functions, a detailed concept is created by the project manager. With these details, a developer implements the OIP. Besides the frontend, the developer also implements rudimentary administrative functionalities in an administration backend. They for instance enable the community managers to monitor the community.

> *First of all, I brief the designer to create mockups. [...] In a second step, we do a briefing where I explain as many points from the concept as possible to the developer, who then develops the platform. (H_PM2)*

The developer uses the HYVE IdeaNet© core system and customizes it to the specifications of the new OIP. Major new functionality is implemented as modules for the HYVE IdeaNet©. It can thus be reused in other OIPs. Smaller adjustments which are particular to the organizer, are implemented directly into the source code and not as a module. Whereas external OIPs are rather standardized, internal ones are often heavily customized (e.g. by an integration into the organizer's corporate IT landscape and processes) and a core asset of HIC:

> *The internal platforms are our asset. That is why we do all of the development in-house. [...] The degree of customization for internal platforms is way higher [than for external platforms]. There are always processes, rights management and so on. The customer has way more modules to pick from and it is always for the long run. (H_CPM)*

The progress of the development is tracked by a feature list, which can be seen as a functional specification. This list contains all required functionality and its status of implementation. The project manager tracks the progress of the development based on the feature list. Despite the wireframes and mockups, the project manager aims at having a first implementation of the OIP as soon as possible in order to discuss details and changes with the organizer based on this draft version.

> *We try to have a first version, like a prototype, as soon as possible to publish it to the client. You can think about a lot of things in theory but the client has to see it in order to exactly tell what he wants. (H_CPM)*

Once all iterations in the build phase are performed, the OIP is ready for deployment.

3.3.4   Deploy phase

The focus of the deployment phase differs for internal and external OIPs. *External OIPs* are mostly hosted at HYVE's infrastructure. Thus, from a technical perspective, deployment is a matter of minutes for go live. From a social perspective, marketing, i.e. community building, starts in order to attract potential innovators. A large part of the effort of external OIP projects goes into the activities of community building and management. This is thus a major task in the deploy phase, although community building for an OIP can already start prior to the technical deployment. HYVE's core competency in this phase is to identify relevant communities to do targeted recruiting of innovators for the OIP[563]. They draw on the skills and techniques of HIR, in most cases the Netnography method.

The deploy phase for *internal OIPs* is more complex from a technical perspective, as the OIP has to be implemented into the organizer's corporate IT landscape in most cases. Therefore, alignment and compliance with the organizer's IT has to be ensured. Connections to existing systems have to be activated and tested. Additionally, while the social perspective of external OIPs is often dominated by marketing decisions, it is more politically driven in internal ones. For internal OIPs, it is important to identify and motivate central actors who can foster the acceptance of the OIP within the organizer's organization. Although there is no blueprint on who to activate and what to do, management support is required:

---

[563] See Bartl and Ivanovic (2010); Fueller et al. (2006); Hueck, Fueller, Bartl and Leckenwalter (2008).

The CEO wrote a letter. That kicked-off the platform and they had 6.000 users on the first day. Subsequently, the middle management pushed the employees to participate, as it would help the overall company. That is a benefit, as the CEO is too far away from most employees. What really counts is that what your line manager tells. (H_CPM)

Besides online activities, there are other approaches to motivate innovators to participate:

One client said that his CEO wrote an e-mail to all employees and pushed the topic. The other one said that he walks from desk to desk and tells each employee the whole story and helps them to submit their first idea. Those are two completely different approaches which both led to success. (H_CPM)

In both cases of internal and external OIPs, the OIP is a web-based tool that is programmed and deployed using a software versioning and revision control system (i.e. GIT[564]) as well as a development (hosted at HYVE) and live version of the OIP (hosted at HYVE or on-site). Due to its centralized architecture in both cases, go live after deployment does not present any problems. In almost all external and most internal OIPs, the go live is on a set and communicated date. Soft launches are rare. This approach is taken to create a strong kick-off with plenty initial ideas and interaction on the OIP. In order to enable a smooth start, the OIP is pre-occupied with content. This content may origin from the project team or some innovators (for instance beta testers from the organizer) that can contribute ideas prior to the go live.

3.3.5 Operate phase

 Once the contests goes live, the work really starts. (H_CPM)

This quote shows the importance of the operate phase in the OIP lifecycle. In order to enable community management, HIC educates the organizers and supports them in community management. Besides updating the status of contributions[565], support has to be given to the community members.

We support a lot at the introduction. Having software is one thing, but if you do not life and manage it, it will not work. You need someone who manages it actively every day as a community manager. (H_CPM)

---

[564] GIT is a distributed version control system; git-scm.com; retrieved September 5, 2012.
[565] See admin links in Figure 38, middle right.

You have to create a viral effect in the company. We manage that. We prepare the newsletters and we are present on the platform. We educate the idea managers on the platform, so they can message the users every day. They evaluate ideas, promise incentives if so and so many ideas are submitted and so on. (H_DEV)[566]

Figure 38: Community management functionalities using the HYVE IdeaNet©

Community management is not limited to actions on the OIP itself. Additional means have to be taken in order to activate the innovators. HIC therefore recommends newsletters and direct communication to raise awareness. Also, banners in the organizer's intranet are an option. The intensity of communication depends on the runtime of the OIP. The shorter the runtime, the more intense is the communication.

You have to write a newsletter to the community every week. Hey hello, this is the innovation community. This is our status. Thanks for being with us. Those who are not in yet, please participate. Those who participate, please submit your ideas. Those who submitted ideas, please evaluate ideas. (H_PM2)

Rules of thumb are based on experience and expert exchange. They help to evaluate the performance of an OIP in order to take additional means if required:

---

[566] See welcome message in Figure 38, bottom left.

By now they have [...] users, hence 10 percent of their employees. That's the average. You have a 90-9-1-rule. 90% [of the users] visit the platform, 9% are active, and 1% are really active. (H_CPM)

Not only those ideas that have a good evaluation are relevant, but in general those, who have a high activity with a lot of comments, visits or evaluations. (H_CPM)

If relevant ideas can be pulled out of the OIP or when the runtime ends, results are documented by the OII. They are presented to and discussed with the organizer or a jury of experts in order to select winning ideas. If possible, HIC supports the organizer with integrating the ideas into its innovation process.

We try to influence the process as much as possible so that there is a wrap-up meeting at the end [of the runtime] where we select the best ideas and create projects from them. Of course, in that point, we reach our limits but we try our very best. (H_PM2)

3.3.6   Optimize phase

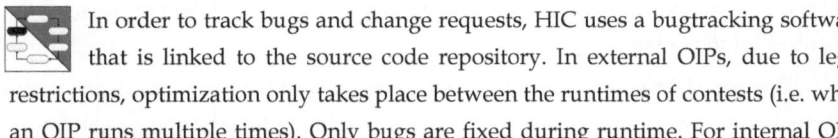 In order to track bugs and change requests, HIC uses a bugtracking software that is linked to the source code repository. In external OIPs, due to legal restrictions, optimization only takes place between the runtimes of contests (i.e. when an OIP runs multiple times). Only bugs are fixed during runtime. For internal OIPs contrarily to external ones, continuous optimization is crucial.

We have frequent releases once the [internal] contest is live because the customer says we now need this and that feature. Then we implement it accordingly. [...] Many customers have the problem that ideas are submitted twice, five times or ten times. In my eyes, that is not a real problem. Quite the contrary: maybe the ideas will be relevant as a cluster! That is why we, for instance, implemented a calculation of similarity for the customer. (H_CPM)

Especially, if the OIP projects are designed for a long runtime, a pilot version might be used at first.

We have many clients that want to start with a small pilot to try the system. The target is to make it a full version after half a year or one year and really customize it at that point. (H_CPM)

Parallel to the integration of research into the best practices of HYVE, HIC integrates organizers to improve the OIPs and processes on the long run. Therefore, a periodic roundtable with organizers is installed to exchange experiences, best practices and problems.

> *What we also do is a biannual user meeting. We invite all customers that are interested. We bring them together to exchange experiences. On the one hand, this exchange is very valuable for the client; on the other hand, we [HIC] get an impression of the bugs and real problems. (H_CPM)*

3.4 Summary and conclusion

The following subsections summarize findings and draw conclusions from the HYVE case concerning the OII (*subsection 3.4.1*), the OIP (*subsection 3.4.2*) and the OIP lifecycle management (*subsection 3.4.3*).

3.4.1 The open innovation intermediary

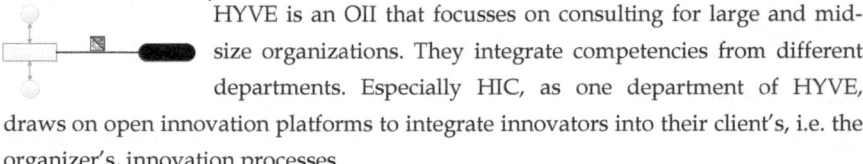 HYVE is an OII that focusses on consulting for large and mid-size organizations. They integrate competencies from different departments. Especially HIC, as one department of HYVE, draws on open innovation platforms to integrate innovators into their client's, i.e. the organizer's, innovation processes.

The organizer is the main player who conducts community management and absorbs knowledge from the community. By the targeted identification and attraction of innovators for an organizer's innovation community, HYVE helps the organizer to draw systematically on a particular type of knowledge. HYVE's approach to identify innovators allows an organizer to rely on individuals who are not yet among their customers. As a consequence, HYVE helps the organizer to identify demands of individuals that have not been focused up to that point.

3.4.2 The open innovation platform

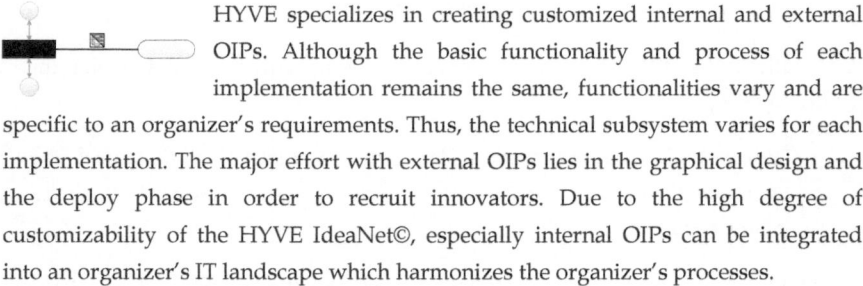 HYVE specializes in creating customized internal and external OIPs. Although the basic functionality and process of each implementation remains the same, functionalities vary and are specific to an organizer's requirements. Thus, the technical subsystem varies for each implementation. The major effort with external OIPs lies in the graphical design and the deploy phase in order to recruit innovators. Due to the high degree of customizability of the HYVE IdeaNet©, especially internal OIPs can be integrated into an organizer's IT landscape which harmonizes the organizer's processes.

The OIP processes HIC usually implements foster single idea submissions. These ideas can be elaborated in comments. Although they have community functionalities, *external OIPs* rather focus on a competitive contest setup than on a collaborative community setup. Innovators are mainly motivated by the opportunity to contribute their ideas and the reward system. In *internal OIPs*, the community spirit dominates the social subsystem. Additionally, the social subsystem of internal OIPs is more complicated than the one of external OIPs as corporate structures have to be considered and, hence, more customizations are required. In both cases, community management is mostly done by the organizer himself who is advised by HYVE's community managers. Thus, the organizer needs some expertise in community management.

From an innovation process perspective, HYVE's approach fosters the search and selection phases. There are no dedicated mechanisms that ensure producibility of community results and acceptance by the organizer. Thus, the organizer has to ensure the proper implementation of the OIP's results into its processes.

3.4.3   The OIP lifecycle management

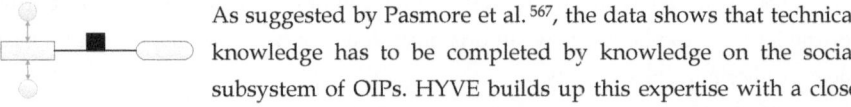

As suggested by Pasmore et al. [567], the data shows that technical knowledge has to be completed by knowledge on the social subsystem of OIPs. HYVE builds up this expertise with a close link to research in the field of open innovation and iterative improvements. An OIP lifecycle manager has to build competencies in managing the social subsystem of OIPs, in addition, or maybe even more than in managing the technical subsystem.

The data shows that the technical implementation of an OIP does not have to be perfect. It rather has to function appropriately. HYVE uses a common basis, namely the HYVE IdeaNet©, which can be seen as a framework for all of its internal and external OIPs. Based on this, they implement a graphical design and custom functionality for each OIP. This results in significant design effort for both, HYVE and the organizer. As a consequence, a diverse codebase exists that requires, on the one hand, a high amount of training for new developers and, on the other hand, plenty of testing activity for a new OIP. This diverse codebase partly owes to a high degree of flexibility and iterative improvements of HYVE, who tries to implement promising ideas for an OIP. On the downside of this, there is a lot of functionality that is specific

---

[567] Pasmore et al. (1982).

to one organizer and hence only used once. HYVE is currently reworking the code to a common basis in order to address this problem. Additionally, simplicity of the code is higher valued than good-practices, like the use of sophisticated frameworks, for the OIP's architecture.

The social subsystem is one reason of this approach to manage the technical subsystem. Due to the low degree of standardized functionality and processes, the social subsystem can be designed very flexible. On the one hand, this provides the freedom to address an organizer's needs, while, on the other hand, requiring a lot of design effort and testing for both, the organizer and the OII. As a consequence, many features of the social subsystem are not reused for other organizers and, hence, only applied once. Organizers benefit from this approach through a lot of flexibility and experience that HYVE gained.

HYVE analyzes its OIPs in order to assess their performance. Many analyses are generated manually in the database, which provides the freedom to derive custom measures. At the downside, it involves manual work and requires that the project manager know how to build corresponding SQL queries.

The organizer is mainly responsible for community management and partly for the jury of experts. Anyhow, an organizer does not have the chance to influence ideas or interim results actively.

HYVE's core competency and focus lies in the attraction of relevant innovators for an OIP. They therefore realize targeted marketing in relevant communities, which are identified based on self-developed methods. Besides this buildup of internal knowledge, HYVE draws on a strong link to the open innovation research community to remain state-of-the-art in designing the social subsystem of OIPs. Standardization at HYVE mainly concerns the overall process on how to design an OIP rather than the OIP design itself. HYVE aims at standardizing workshops with organizers in order to cover all relevant topics and information that are needed to design an OIP.

In short, scalability is limited at HYVE, but they master and enjoy exploring new functionalities for an OIP. As a consequence, organizers have to invest time in order to run a custom OIP project. This applies to for the design as well as to the management of the OIP.

According to the aspects presented above, HYVE focuses on the design of the social subsystem and the deploy phase. An organizer has to hold at least some expertise in open innovation to carry out the community management although HYVE trains them.

From a strategic management perspective, it might be interesting to check whether HYVE has a viable business model that is hard to copy. Under this aspect, two issues remain to be solved: Firstly, how proprietary and volatile is HYVE's knowledge? Secondly, is it possible for agencies who want to compete in the field of OIPs to build up this knowledge easily or not? This applies accordingly to the technical realization of OIPs. HYVE heavily customizes OIPs. It is doubtful which results can be realized with a standard (though configurable) and, thus, less expensive, OIP. Table 22 summarizes the key findings of the HYVE case.

Table 22: Key findings of the HYVE case

	OIP-LM focus	Design, deploy and operate, though build is required for each OIP project
	Key findings	Focus on the design of the social subsystem and OIP deployment
		Social subsystem designed according to organizer's needs
		Targeted attraction of innovators
		Most resources invested in the build phase of an OIP
		Different approaches to internal and external OIPs
		Custom OIP projects for each organizer with very limited scalability
		Almost no out-of-the-box functionality
		Community management done by organizer and supported by HYVE
		Self-perception: innovation consultancy, not a provider of technology

4 Case 3: Atizo

The third case investigates the open innovation intermediary Atizo with its OIP atizo.com. Table 23 summarizes the key facts on the Atizo case. The structure of this case follows the structure described in chapter 2.

Table 23: Key facts on the Atizo case

OII	Atizo AG (referenced as Atizo)
Description	innovation agency with a standardized OIP that runs multiple internal and external innovation projects on the same OIP and in a default or custom graphical design
Foundation	2007
Employees	12 (including 2 founders)
Homepage	www.atizo.com
OIP	www.atizo.com
Tools	innovation community; innovation contest; innovation marketplace

4.1 The open innovation intermediary

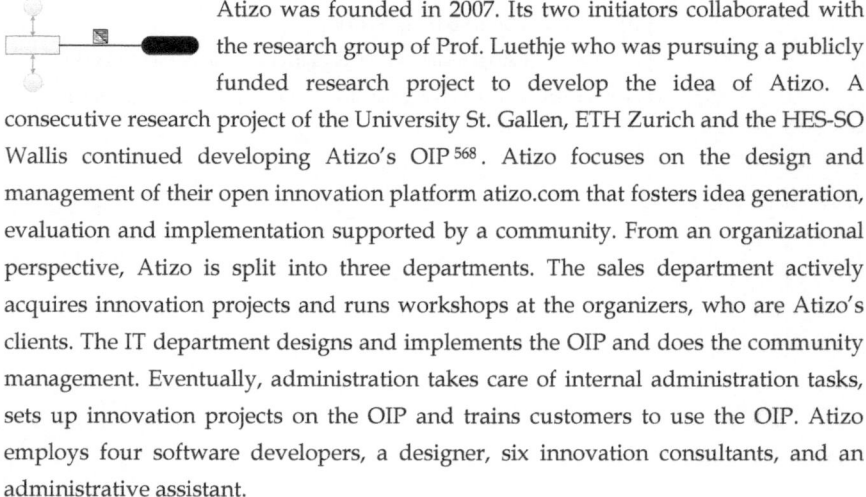

Atizo was founded in 2007. Its two initiators collaborated with the research group of Prof. Luethje who was pursuing a publicly funded research project to develop the idea of Atizo. A consecutive research project of the University St. Gallen, ETH Zurich and the HES-SO Wallis continued developing Atizo's OIP [568]. Atizo focuses on the design and management of their open innovation platform atizo.com that fosters idea generation, evaluation and implementation supported by a community. From an organizational perspective, Atizo is split into three departments. The sales department actively acquires innovation projects and runs workshops at the organizers, who are Atizo's clients. The IT department designs and implements the OIP and does the community management. Eventually, administration takes care of internal administration tasks, sets up innovation projects on the OIP and trains customers to use the OIP. Atizo employs four software developers, a designer, six innovation consultants, and an administrative assistant.

---

[568] Hirsig and Hirschmann (2010).

Besides the traditional way of the sales department to drive sales, Atizo addresses mid-size consultancies to use Atizo in their consulting projects[569]. Most of Atizo's customers, such as Zurich, Rivella, Wander, O² and Mammut, are based in Switzerland. The community mainly spans German-speaking countries. The following section addressed Atizo's OIP Atizo.com.

4.2 The open innovation platform

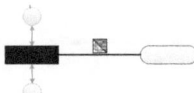

The following two subsections introduce the OIP Atizo.com (*subsection 4.2.1*) as well as its processes, and core functionalities (*subsection 4.2.2*).

4.2.1 Overview

Atizo's community comprises more than 15.000 innovators from a general audience and a diverse set of disciplines and ages. The innovators have already generated more than 73.000 ideas in about 180 innovation projects[570]. Most innovation projects on Atizo address innovation in the area of consumer goods, insurances and services.

Atizo provides three modes of its software-as-a-service OIP that offers to run both, standard and custom OIP projects. Firstly, using *Atizo Private*, organizers can start brainstorming projects on Atizo for free. This mode supports only basic functionality to brainstorm and discuss ideas and does not include any support, custom design or evaluation of ideas. Secondly, acquiring *Atizo Business*, an organizer can use the full functionality of the OIP to start invite-only innovation projects, for instance to innovate with employees or suppliers. Thirdly, *Atizo Community* allows the integration of Atizo's community of innovators in order to support an organizer's innovation project. Using Atizo Business or Atizo Community, the organizer can select a set of service modules to support its innovation project. These modules include a custom graphical design of the innovation project, an advanced user management, support, training, and workshops[571]. A list of major projects can be found in Table 24.

The OII adopts multiple roles, according to the purchased modules: Atizo helps organizers formulating the innovation problem. They also offer an OIP and processes in order to work on the innovation task collaboratively. Furthermore, they

---

[569] According to Atizo, mid-size consultancies employ about 10 to 100 consultants.
[570] As of August 2012. For statistics see www.atizo.com/products/community/; retrieved August 20, 2012.
[571] For a list of prizes and a description of the modules see business.atizo.com/products/; retrieved August 20, 2012.

support or train organizers to setup and manage an innovation project using Atizo's OIP. The details of the task division between the organizer, the innovators, and the OII are described below. A project carried out for Rivella provides a suitable example to illustrate the processes of Atizo[572].

Table 24: Selected projects on Atizo (in chronological order)

Organizer	Project	Duration	Ideas / Evaluations	Result
Bischofszell	Innovative soft drinks	50 days	475 / 912	New bio-ice tea Glueckstee
BMW	Motor-bike of the future	103 days	769 / 1224	Several concept-models for future development
KPT/ CPT	Health-file online	41 days	291 / 559	Online-reminder for inoculations
Bell	New products made of Swiss chicken meat	59 days	504 / 968	New products Pouletburger and Poulet Meatballs
K Kiosk / Valora AG	Kiosk of the future	54 days	626 / 1291	Several products and concepts that influenced the main concept of a kiosk
Axa	Insurance cooperation partner	21 days	341 / 655	The winning idea confirmed internal plans and influenced the solution
Midor AG	Products by customers	16 days	429 / 900	Snack made of healthy grain kinds
Rivella	Search for new flavor	32 days	801 / 1538	The best ideas will be implemented and evaluated by the community: this will define the new product

4.2.2 Processes and functionality

The Rivella AG is a Swiss beverage company that uses the Atizo Community version process, as shown in Figure 39, to develop line extensions for their lemonade Rivella. The lemonade is the second most sold soft drink in Switzerland behind Coca Cola[573].

---

[572] See www.atizo.com/projects/ideas/1599/gesucht-das-neue-rivella/; retrieved August 24, 2012.
[573] See www.rivella.com/ch/index.php?id=101&L=0&tx_indexedsearch%5Bsword%5D=kennzahlen; retrieved August 24, 2012.

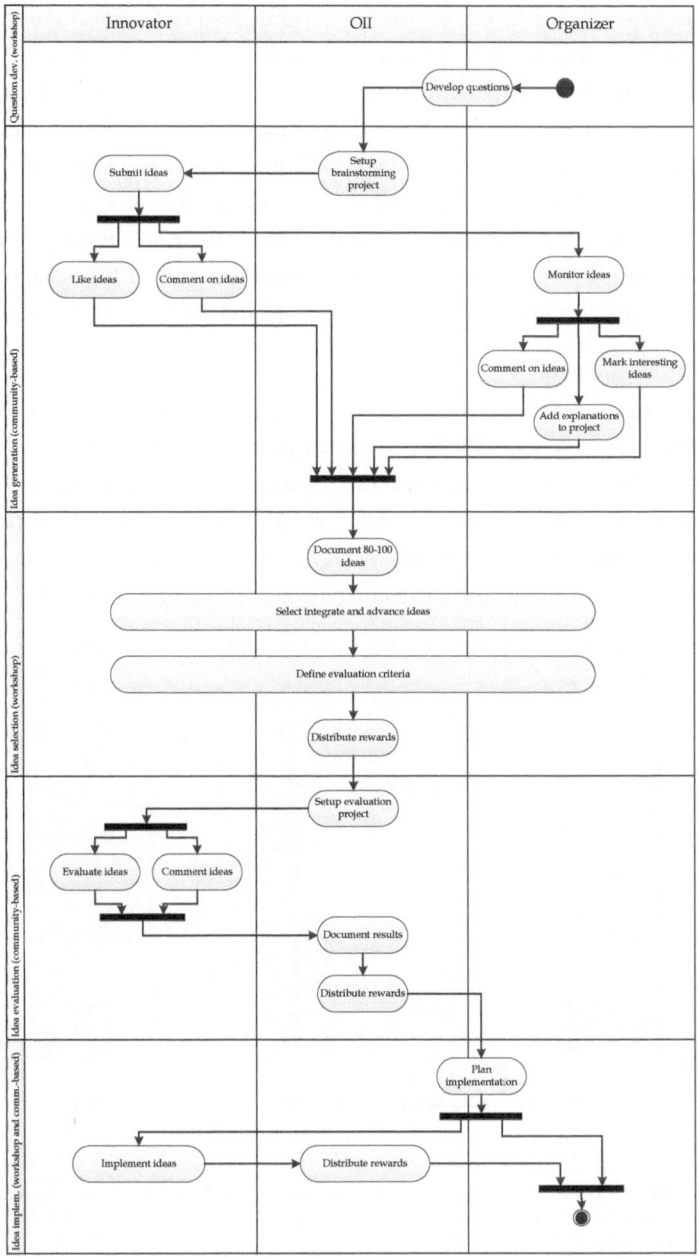

Figure 39: Activities in an Atizo Community project

Rivella uses the whole potential of Atizo to integrate their consumers in the search for new products, their selection, and implementation. Atizo's process contains five steps in realizing the integration of the innovators, namely *question development, idea generation, idea selection, idea evaluation* and *idea implementation*. Each step that involves the innovators (idea generation, idea evaluation, and idea implementation) is conducted in a separate workspace of the OIP. Organizers can opt to select only single steps of the process or they can run throughout the whole process. They could run, for instance, an evaluation project of existing ideas only, if they wanted to. The five steps of Atizo's process are shown in Figure 39 and explained in the following.

Question development

In the first step, Atizo runs a two-hour workshop in collaboration with the organizer to generate appropriate questions and sketch evaluation criteria. For this purpose, Atizo firstly brainstorms the innovation project with the organizer and roughly prioritizes the outcome. With this input on the intention of the innovation project, the innovation consultant of Atizo proposes three questions the innovators can work on. They are discussed with the organizer and merged into a single question[574].

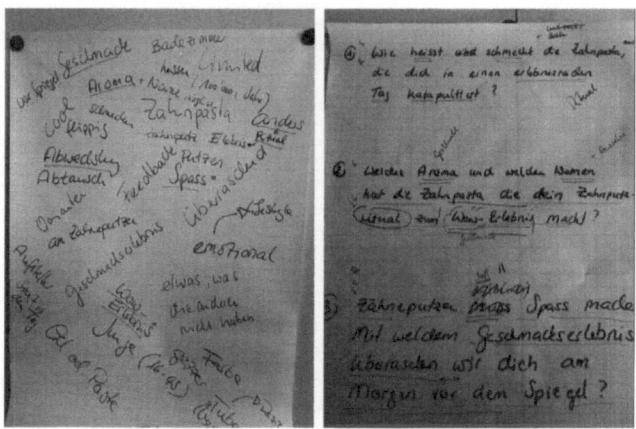

Figure 40: Brainstorming and question development in an Atizo workshop[575]

---

[574] See Figure 40. Green marks parts of the three proposed questions (or respectively the brainstorming) that the client likes and red marks parts that the client does not like. The final question is composed of the green parts.

[575] These are sample photographs of flipcharts from a workshop that was not conducted for the Rivella project.

The final part of the workshop is devoted to defining the title of the innovation project and rough evaluation criteria. The latter provide some guidance for the innovators concerning the intended outcome of the innovation project. Additionally, the communication strategy is determined, i.e. content and channels of communication as well as the persons involved.

Idea generation

In this phase, the community generates ideas for the published question. For this purpose, Atizo creates a brainstorming project on their OIP where innovators can submit ideas, discuss them and like their favorite ones[576]. The organizer comments ideas, marks interesting ones and might add further details to the innovation project description if necessary. This phase usually runs between four and six weeks.

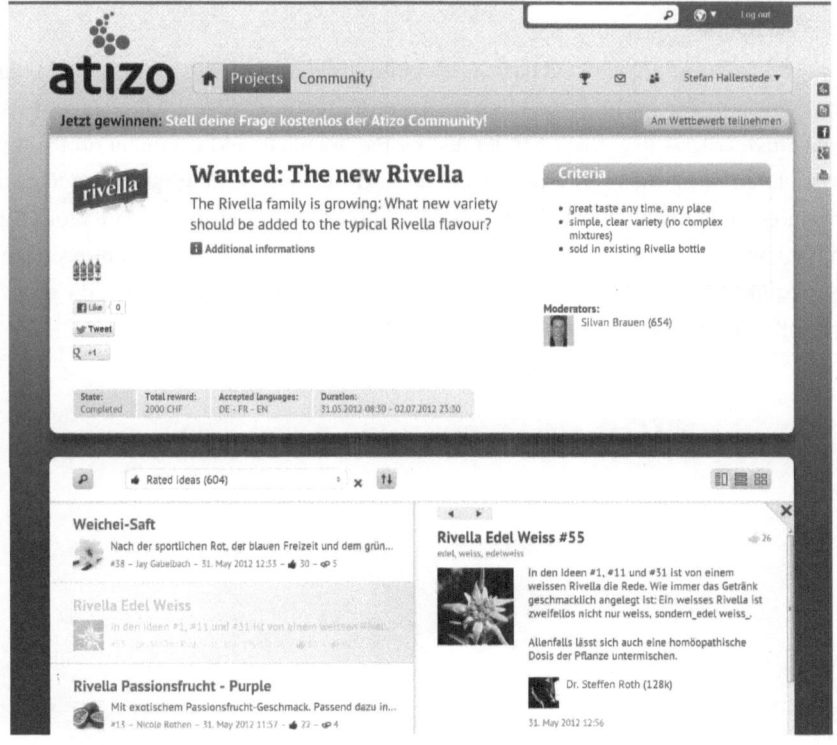

Figure 41: Idea generation project on Atizo

---

[576] See Figure 41.

Idea selection

The best 80-100 ideas according to the organizer's (interesting ideas) and innovator's (likes and comments) perception serve as input for an idea selection workshop that takes half a day. In this workshop representatives of the organizer coact with innovators in order to consolidate the 80-100 ideas to 10-15 elaborated ideas. An innovation consultant of Atizo moderates the workshop, which starts with a clustering of the unstructured 80-100 ideas. Similar or complementing ideas are integrated and advanced to elaborated ones. The elaborated ideas are developed by the workshop participants and described according to a scheme, called idea outline. By doing so, an idea outline is created of each of the 10-15 elaborated ideas which are the output of this workshop. All ideas from innovators that inspired an elaborated idea are rewarded by the OII with a share of the overall price money.

Idea evaluation

The 10-15 idea outlines are fed back into an idea evaluation project on the OIP. In this, the innovators can evaluate the ideas according to quantitative (e.g. Likert scales) and qualitative criteria (e.g. free text) defined by the organizer and comment on them in order to advance them. The organizer can also define screening questions like for instance "How often do you drink Rivella?" in order to allow an evaluation of ideas only by innovators who fulfill certain criteria. In this sample evaluation project, they are required to have experience with Rivella's beverages[577]. An evaluation project typically runs for one week.

Figure 42: Screening questions in an Atizo evaluation project

---

[577] See Figure 42.

Idea implementation

The results of an evaluation phase are documented, statistically analyzed and presented to the organizer using a structured template. If requested, Atizo runs a two-hour implementation workshop with the organizer. In this workshop, all ideas that have been evaluated are discussed and promising ones are selected for implementation. An owner of the idea and timelines for the implementation are defined. Figure 43 shows the outline of an idea in the implementation workshop including results from the evaluation as well as a roadmap for the implementation of ideas. If required, an innovation project can be set up on the OIP to request implementation proposals from the innovators or to recruit innovators for the implementation. The following section addresses the lifecycle management for the OIP that has been introduced in the present section.

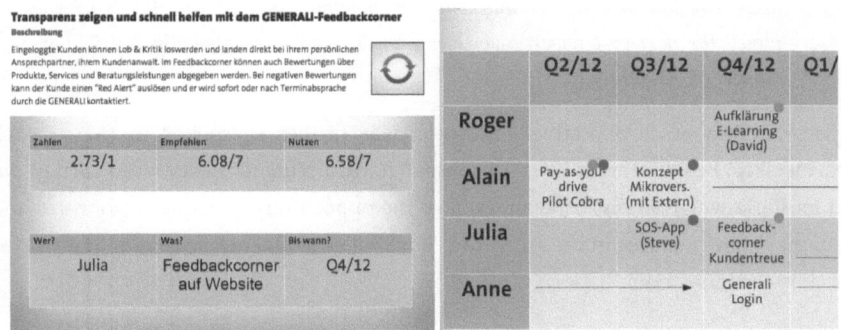

Figure 43: Idea outline and implementation roadmap in an Atizo workshop[578]

4.3   The OIP lifecycle management

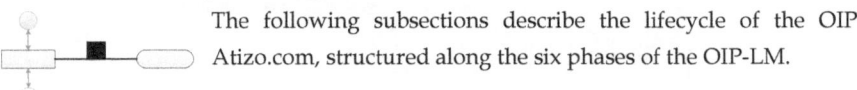

The following subsections describe the lifecycle of the OIP Atizo.com, structured along the six phases of the OIP-LM.

4.3.1   Requirements phase

The idea of Atizo's OIP originates from observations of the activity on Wikipedia. The founders originally intended to write down knowledge and make it accessible and acquirable to the world. This idea, paired with experience from

---

[578] This is a sample documentation from an implementation workshop that does not origin from the Rivella project.

the open source world, led to the idea of a problem-oriented OIP. The major aim is integrating diverse knowledge to the highest possible degree.

> *We want to keep it as diverse as possible. Everybody should be able to participate because, ideas to not necessarily origin from one sector. Often real innovations stem from another sector you did not bear in mind. (A_CEO)*

With a first concept in mind, Atizo works closely with researchers from the field of open innovation in order to profit from latest results of research. The researchers help defining requirements for the OIP like in this example:

> *We needed some kind of acknowledgement for the users. [...] We also found that in studies: The best would be the company joining a close dialogue with the community, but companies are not up to that. They do not want to explain for 500 ideas why they like the one any why not the other one [...] and that is why we picked the next best reward, which is divisible and internationally distributable: money. (A_CTO)*

Also the idea of collaborative innovation was further sharpened in the research partnership. Besides this close link to research, requirements elicitation is carried out in multiple ways. Firstly, innovators have the opportunity to request, prioritize and comment new features in the Atizo idea box[579]. The status of a feature request is fed back to the innovators.

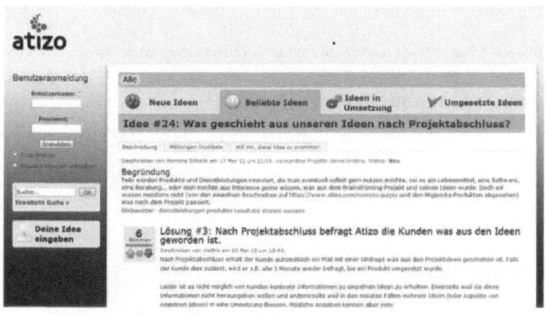

Figure 44: Atizo idea box for feature requests of the innovators[580]

---

[579] See Figure 44.
[580] ideabox.atizo.com; retrieved August 31, 2012.

Secondly, especially in the course of the requirements elicitation for the initial version of the OIP, a prototype was published. Representatives from 15 companies and about 500 students from the research partners tested and evaluated features in order to come up with new requirements. Thirdly, discussions with organizers revealed that, despite of having the capability to run a standalone version of the OIP, organizers prefer to run a software-as-a-service version due to rigid IT departments. If a requirement is specific to an organizer and not usable for others, the requesting organizer is charged with costs for this development. Fourthly, Atizo exploits experiences they gain. The original idea of the OII, for instance, did not include any consultancy, but only the OIP, which was meant to sell itself online. Atizo figured out that this did not work out and started to build up a sales department and an accompanying consultancy process to set up innovation projects. In the same iterative improvement approach, the organizer-innovator integration arose:

> *We recognized that we need a form of feedback so that the company can engage with the community. We implemented that, tested it and now it works quite well. (A_CTO)*

Atizo assigns areas of competence, like for instance the area of community functionalities, to its developers. A developer is responsible to gather and manage requirements within the assigned area of competence, no matter where a requirement origins. Once the developer has gathered sufficient requirements for an area, specifications are created and realized as described in the following subsection. Requirements are prioritized according to a pre-defined share of development time and the organizers' and innovators' priorities:

> *We want to use a particular amount of time for certain topics. We want to spend time for the community, for the project moderation, for our internal processes and so on. That is why we assign time shares for each of these topics and one developer is then responsible for one topic. He searches for inputs on the most pressing topics. (A_CTO)*

4.3.2 Design phase

Also in the design phase Atizo draws on research partnerships in particular to design the social subsystem of their OIP. The focus lies on facilitating appreciation among innovators on the OIP and from the organizer while trying to keep the need for monetary rewards as low as possible. In order to realize this vision, Atizo works closely with its research partners. Especially the underlying five step process of an Atizo project[581] is designed in cooperation with researchers, who also publish results from the cooperation with Atizo[582]. Details of Atizo's process are defined in a perpetual-beta approach. Accordingly, new features are designed in incremental steps with frequent feedback from stakeholders. If a feature is requested by a particular individual, it is designed in cooperation with this particular individual.

> *It was more like having a vision. Someone posts a question and another one has to be able to answer it. Possibly incorporate some collaboration. And then we built it and searched for companies that can use it. (A_CTO)*

But not all organizer requests are realized, as the following example shows:

> *We actually offer it as software-as-a-service. In principle, we could do it [the implementation] on-site, but so far, we convinced all of our clients that we do not have to do that. (A_CTO)*

Accordingly, Atizo uses its experience from the field to guide organizers. As described in subsection 4.2.2, Atizo applies a multi-step process that draws on workshops to define an appropriate question and to prepare an evaluation project. These workshops apply a structured and standardized approach to address the tasks of designing an innovation project on the OIP while simultaneously guiding the organizer.

> *We recognized that the companies have difficulties formulating good questions. They tend to post too long questions nobody wants to read. The second problem is a language nobody understands. Due to these two problems we support our clients in formulating a question. [...] We are leading edge at this. We do it for four years by now and did by far the most projects. I think we did more than 150 projects with our process. (A_CEO)*

If the organizer plans to work with Atizo on a frequent basis, Atizo offers training to instruct the organizer on how to run the question development workshops on their own.

From a software architecture perspective, the OIP has to support the multiple options Atizo offers, like for instance an internal or a community-based project. In

---

[581] See subsection 4.2.2.
[582] See for instance Gassmann (2010).

order to facilitate reusability, Atizo incorporates multitenancy to its architecture. In Atizo's OIP architecture, there are standard modules that can be overwritten to adapt a tenant of the OIP to an organizer's needs:

> *All of it is already there. Then we have an app-structure: If a company wants a custom idea submission form, we build an app for that. This app overwrites the standard functionality. (A_CTO)*

Minor adaptions, like the adjustment of the color scheme, group management, targeted innovator selection and advanced statistics can be done from an administration backend without the need for a separate tenant. Frequently used adaptions are implemented in an administration backend.

> *It depends. I think at some point it does not make sense any more to build a backend. There is no need to be able to configure everything in a backend. Even if you try, there is going to be the next client who has a special wish that requires a custom implementation. (A_CTO)*

Owing to this architecture, the basic structure of the code can remain unchanged while fulfilling an organizer's needs. Besides multitenancy, Atizo supports multiple languages, i.e. English, French and German. They draw on a partly object-oriented programming paradigm on the basis of Jumbo Website Manager[583], which serves as a content management system and framework at the same time. The architecture realizes a model view controller architecture using PHP and Phyton[584]. Website tracking is realized with Piwik. Atizo uses open source software only. The OIP is hosted on a self-managed LAMP[585] stack.

4.3.3   Build phase

 Atizo explicitly organizes its development according to Scrum. According to the developers, they need an agile approach that allows them to quickly react to new or changing requirements. Thus, outsourcing, although possible, is not an option. The backlog of a sprint is defined by a single developer while the implementation is carried out by the whole development team:

---

[583] Jumbo Website Manager is a content management system for websites using PHP and JavaScript; modset.net; retrieved August 21, 2012.
[584] Phyton is a programming language that is built to facilitate the integration of modules from different programming languages; www.python.org; retrieved August 22, 2012.
[585] LAMP is an acronym for Linux, apache, MySQL and PHP (or Perl/ Phyton) which denotes the used technologies in a web server. See Gerner, Naramore, Owens and Warden (2006).

> *We have tasks and stories. We estimate them and have a roadmap to plan our development. That is how we coordinate us. [...] The developer is only responsible for enough requirements in his topic. We implement it all together, dedicating us to a single topic at a time. (A_CTO)*

This approach helps Atizo to finish necessary functionality as quickly as possible. In order to meet high quality standards, Atizo employs four developers, two of which have studied informatics while the other two have had vocational training in that area. In addition to the testing of the developers, Atizo draws on external beta testers to prevent the go live of bugs and ensure usability.

> *It is not anonymous. There are many registered users we can ask. In addition, we have a group of beta testers. They are very committed and most of the time, they answer our questions very easily and precisely. Results of them are more certain than when we do any random statistics of traffic data, which is often hard to interpret. (A_CTO)*

The beta testers can access a development environment. The focus on usability of the OIP results in substantial testing and feedback circles. Thus, the actual effort to implement new functionality depends on the intended functionality:

> *Often realization is the most difficult part but that can change rapidly. If you have to integrate externals in planning and usability-testing, it is way more effort than the actual implementation. But if it is technically complex but easily describable, it is the opposite way around. Thus, it depends heavily on the case. (A_CTO)*

In order to keep the number of published bugs as low as possible, Atizo additionally draws on automated tests.

> *We have a very detailed coverage of automated tests. [...] There are not many bugs that go live. If we have a bug in the live system, it is mostly very hidden, for instance in an administrative function, actually things almost nobody uses. (A_CTO)*

The OIP's source code is managed using the versioning system GIT.

4.3.4   Deploy phase

The Atizo team deploys on average twice a day. They thus react quickly on changed requirements or bugs. Due to the use of a repository (i.e. GIT) and their SaaS architecture, deployment of new functionality is easy from a technical perspective. Also innovation projects can be deployed to the innovators using the administration backend functionality. The innovation consultants provide the required information, i.e. the innovation project title, the questions, evaluation criteria etc., which were gathered in the workshops. The setup of an innovation project itself is done by administration.

Besides the technical deployment, Atizo helps the organizer with the social aspects of deployment, such as marketing. Multiple strategies are applied. Firstly, an organizer can opt to be featured in a newsletter to the innovators and to have a prominent position on the homepage. In addition, other social media marketing channels and internal marketing at the organizer are exploited.

> *Rivella is a good example for that. They wanted to build their own community from their Facebook fans. They started with a public project and only then users registered with their custom platform. [...] Now users are active on both, Atizo and Rivella's platform. (A_ADM)*

In addition to marketing, the community managers from the organizers receive training on what to do and how to interact with the innovators.

> *I [the administrator] inform them what they exactly have to do. We do the introduction by telephone. That is 15 minutes where I show them the system, the statistics and how they mark interesting ideas. (A_ADM)*

If an organizer has not booked any workshop or training, they can still receive short introductions to the basics of developing a question and community management via how-tos to get into the topic[586].

4.3.5 Operate phase

 During the runtime of an innovation project, Atizo monitors the innovators and helps both, innovators and organizers, with administrative questions.

> *Once the project is online, I do not do very much. It is mostly the community that is active. I ensure that the questions of the community are answered and that the moderator of a project addresses issues of the community. The moderator has to guide the community in case ideas shift towards the wrong direction. (A_ADM)*

---

[586] See www.atizo.com/instructions; retrieved August 29, 2012.

Mostly, questions are answered by the development team in order to obtain direct feedback for the developers. Due to the reciprocal character of the community, fraud is usually not a serious problem. The innovators rather take care of it themselves.

> *We cannot monitor everything, but if the community or the moderator reports something, we will have a look if our general terms and conditions are violated. If that is the case, we delete the contribution and inform the contributor. Depending on the type of violation, we reprimand him and if that happens again, we kick him out of the community. [...] The community governs itself very well. In case of abuse, there are quickly quite a lot of comments. (A_ADM)*

This way, communication takes place via multiple channels, directly on the OIP, via email or telephone as well as in forums and blogs. From a content perspective, the organizer takes care of community management, answers questions, marks interesting ideas and adds additional information to an innovation project if needed for clarification of the task. Atizo recommends devoting one to three community managers on the organizer side to a project. More community managers would call for too much coordination effort. In any case, only one community manager is visible to the community to represent a single point of contact. This distribution of responsibilities between the OII and the organizer is installed intentionally:

> *We discussed it quite often internally how to automate that interesting ideas advance into the workshops. That is extremely difficult. We cannot really put ourselves in the company's position because we do not exactly know what they can and what they want to realize. [...] That is why we depend on the company's community management. (A_ADM)*

By this, the organizer can direct the innovators towards a desired outcome, for instance towards products that are producible for the organizer. In addition to this, the innovators receive appreciation through the organizer's feedback, which Atizo sees as a crucial success factor for an innovation project on their OIP.

> *Although there is no contact between the company and community at great length [referring to commenting every single idea], we try to keep the contact as close as possible in order to express the appreciation. For this reason, the company has to do the community management and mark interesting ideas itself. (A_CEO)*

> *During the runtime of a project I check at least once with the client how it works. I contain if they are really active or just wait for the ideas to come in. I advise them*

to write comments and mark interesting ideas so that the community gets a feedback. (A_ADM)

By the continuous interaction with the innovators and especially the identification of interesting ideas by the organizer, the idea selection workshop is facilitated. 400-800 ideas are generated in a typical idea generation phase. The OIP automatically generates a report from the best 80-100 ideas in a format that serve as a basis for the idea selection workshop. In general, Atizo estimates that about 20% of the generated ideas are usable. Analogous, the reports from an evaluation phase are generated automatically:

They are directly generated as a spread sheet. Everything is automated with graphics and means so that the client can use it. (A_ADM)

The appreciation for the innovators is furthermore strengthened by offline events. Atizo organizes community meetings on a regular basis that do not have a particular purpose other than to get to know each other. In addition, as described in subsection 4.2.2, innovators are invited to idea selection workshops with the organizer.

An idea implementation workshop that follows an ideation or evaluation project is designed to foster idea implementation at the organizer, which is a major concern of Atizo. In this workshop, the organizer's employees oblige each other to the implementation. Results of the implementation workshop are transferred back to the innovators. The success of an idea's final implementation depends on the whole process Atizo applies.

In the beginning, we had projects that did not result in a product [because the process was not yet optimized]. But for instance with Migros, we now have many things that are realized mainly owing to a good process. (A_CEO)

Atizo monitors the progress of an idea implementation and market introduction. If an idea is finally realized, Atizo writes short stories of success in order to document a project and provide feedback for the innovators[587]. In order to keep innovators up-to-date, newsletters are sent to all innovators or specifically to innovators of a particular innovation project on a frequent basis. After a successful ideation, evaluation or implementation phase, rewards are distributed by the OII.

---

[587] See www.atizo.com/success-page/; retrieved August 29, 2012.

4.3.6   Optimize phase

 Atizo continuously optimizes its OIP as the average of two source code commits per day shows. Input for these optimizations is manifold. However, the major source is feedback from the innovators, which origins either from the idea box or from direct feedback. Website tracking is only used if necessary, as the figures are easily misleading and misinterpreted for the purposes of Atizo. Additionally, tracking reduces the performance of the OIP. Like in the design phase, optimization draws on results of ongoing research projects.

> *We are currently working on a project to automatically group short texts. We have first promising results that we will integrate in the next months to see if that is really useful. (A_CTO)*

Feedback from the organizers is gathered using a structured feedback form[588]. In doing so, also the success of an innovation project is assessed from the organizer's perspective.

Figure 45: Organizer feedback form of Atizo[589]

4.4   Summary and conclusion

The following subsections summarize findings and draw conclusions from the Atizo case concerning the OII (*subsection 4.4.1*), the OIP (*subsection 4.4.2*) and the OIP lifecycle management (*subsection 4.4.3*).

---

[588] See Figure 45.
[589] www.atizo.com/survey/set/7/; retrieved August 29, 2012.

4.4.1   The open innovation intermediary

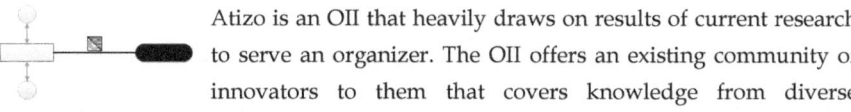

Atizo is an OII that heavily draws on results of current research to serve an organizer. The OII offers an existing community of innovators to them that covers knowledge from diverse domains. By means of this offer, the OII builds bridges to allow the absorption of new ideas by an organizer. By running multiple concurrent innovation projects that incorporate innovation contests on one OIP, the OII offers traits of brokering innovations. As consumers are among Atizo's innovators as well, their demands are articulated, which provides valuable insights for an organizer.

In its standardized process to run OIP projects, the OII supports the organizer in processing the innovator's results. Therefore, workshops are put in place to integrate and improve contributions by the innovators and to plan their implementation.

4.4.2   The open innovation platform

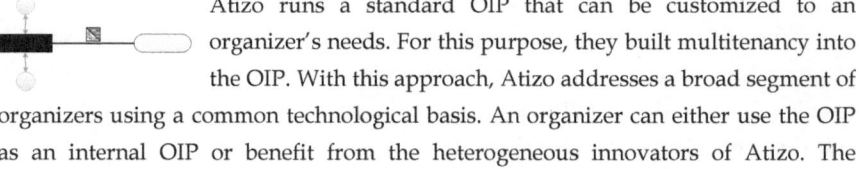

Atizo runs a standard OIP that can be customized to an organizer's needs. For this purpose, they built multitenancy into the OIP. With this approach, Atizo addresses a broad segment of organizers using a common technological basis. An organizer can either use the OIP as an internal OIP or benefit from the heterogeneous innovators of Atizo. The modular structure of the service packages, among which an organizer can select, allows for a custom, but yet transparent OIP project that might serve different phases in the innovation process. Although, the OIP and its accompanying processes represent all phases of the innovation process, Atizo is especially strong in the search and selection phases, which are represented by ideation and evaluation projects on the OIP. During the implementation phase, Atizo mainly builds on offline workshops.

Through the process Atizo applies, the innovators focus on the relevant task of a phase, i.e. ideation, evaluation or implementation. Especially the evaluation phase in an innovation project is interesting. Evaluation can be allowed for a subset of the innovators by applying screening questions. Thus, an organizer can ask for input from innovators with certain characteristics and consequently perform a targeted evaluation.

The OIP itself is strongly integrated in the offline activities. Even innovators might be integrated into the different workshops, although this only applies to approximately five innovators per project. The input for a workshop is automatically generated from the OIP. The main aim of this online-offline-integration is to overcome the NIH syndrome as well as to improve the outcome, by integrating the organizer into the innovation project. Thus, the whole OIP is targeted at products and services that can and will be implemented by the organizer.

The community that addresses and comprises a broad audience of innovators, is thereby built on mainly non-monetary rewards like the joy of innovating and acknowledgement from peers and the organizer. To facilitate this, the OIP asks for feedback from innovators as well as from organizers. In addition, the communicated and sought aim of Atizo is to generate output that is marketed by the organizer. This output is tracked and fed back to the innovators actively in order to address motives of the innovators.

4.4.3   The OIP lifecycle management

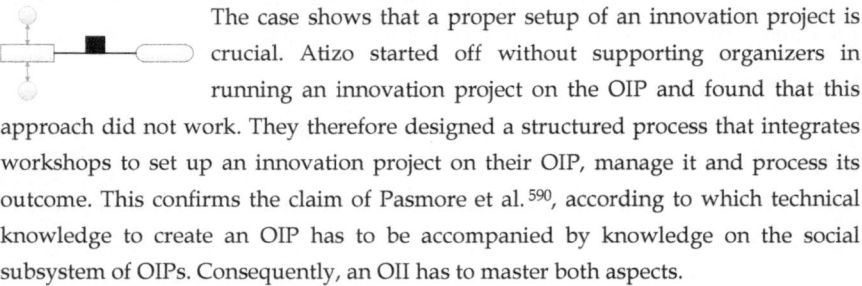 The case shows that a proper setup of an innovation project is crucial. Atizo started off without supporting organizers in running an innovation project on the OIP and found that this approach did not work. They therefore designed a structured process that integrates workshops to set up an innovation project on their OIP, manage it and process its outcome. This confirms the claim of Pasmore et al.[590], according to which technical knowledge to create an OIP has to be accompanied by knowledge on the social subsystem of OIPs. Consequently, an OII has to master both aspects.

In order to establish the required expertise, Atizo builds on their experience from open source projects and heavily integrates researchers and research projects in the field of open innovation into their OIP design activities. Beyond that, details of the social subsystem are designed using a perpetual-beta approach. The technical subsystem is addressed by experienced developers, who apply state of the art principles for software design. The interesting finding here is that Atizo applies a content management system, Jumbo Website Manager, as a basis for their OIP. They can thus quickly create new functionalities of their OIP while still having the freedom to add custom functionality, as Jumbo Website Manager serves as a content management system and framework at the same time. Due to the multitenancy and modular architecture of the OIP, all organizers can benefit from developments while requests that are specific to an organizer can be handled without compromising the architecture as well.

---

[590] Pasmore et al. (1982).

Atizo's core competency is their process and the workshops they moderate to set up an innovation project or to process the results. Besides the structured approach, Atizo helps organizers to select the right individuals for a workshop, recruiting from both, the innovators and the organizer.

Atizo strives to delegate as many aspects of community management to the organizer as possible, while simultaneously providing guidance on what to do. Therefore, trainings are done and documentation is placed to the organizer's disposal. If an organizer runs innovation projects using Atizo frequently, also training in setting up a project is provided. Thus, Atizo helps organizers to manage OIP projects themselves and chooses to monitor the performance. Due to the organizer's mainly indirect influence when interacting with the innovators (for instance by marking interesting ideas), this approach is viable, as only little damage can be caused by the inexperienced organizer when dealing with the innovators. Besides the delegation of the community management to the organizer, also the documentation of results that the innovators create is automated. Thus, only little effort for Atizo is required to run an innovation project. In sum, Atizo focuses on the informed design of their OIP and standardizes the processes to operate it. Table 25 summarizes the key findings of the Atizo case. In chapter 5, the present part, including the cases that were introduced, is briefly summarized.

Table 25: Key findings of the Atizo case

OIP-LM focus	Design and operate
Key findings	Design of the OIP informed by and developed in cooperation with research
	Focus on the social subsystem
	Modules with services by the OII can be freely combined
	Community management done by organizer and supported by Atizo
	Evaluation by targeted audience possible
	State-of-the-art technical subsystem
	Multitenancy allows for custom or standard, internal and community based OIPs while exploiting synergies across innovation projects
	Self-perception: OIP provider with supporting innovation consultancy

5 Summary of Part III

This part introduced the empirical findings that are used to elaborate the OIP lifecycle model. A multiple in-depth case study design following a linear-analytic structure according to Yin[591] presents the research method of choice. Three cases were assessed in this setting using semi-structured qualitative explorative interviews as the main data source. The selected case subjects, namely the innosabi GmbH, the HYVE AG and the Atizo AG, are established open innovation intermediaries that frequently design and manage OIPs. The cases were selected using an information-oriented approach to represent maximum variation. In this line of argumentation, the IT-based tools for open innovation, as introduced in Part II, are covered by the OIPs of the OIIs. For each case, the OII, the OIP, and the approach to design and manage the OIP were outlined. The latter was structured along the six phases of the OIP lifecycle model. Table 26 summarizes the key findings from the three cases. The discussion of the presented results and a cross-case analysis follows in Part IV.

Table 26: Summary of key findings from the three cases on OIP lifecycle management

OII	innosabi GmbH	HYVE AG	Atizo AG
OIP	unserAller.de	HYVE IdeaNet©; e.g. gemeinsamselten.de	atizo.com
OIP-LM focus	Operate and optimize	Design and operate	Design and operate
Key findings	Focus on the social subsys.	Focus on the social subsys.	Focus on the social subsys.
	Most resources invested in community management	Most resources invested in building individual OIPs	Most resources invested in development and sales
	State-of-the-art technical subsystem	Different approaches for internal and external OIPs	State-of-the-art technical subsystem; multitenancy
	Standard OIP	Custom OIP for each org.	Configurable OIP
	Sophisticated administration backend	Community management done by organizer	Community management done by organizer
	Self-perception: provider of technology with open innovation consultancy	Self-perception: innovation consultancy, not a provider of technology	Self-perception: provider of technology with open innovation consultancy

---

[591] Yin (2009).

Part IV

Discussion

1 Functions of open innovation intermediaries

Part III introduced the findings from three cases on OIP lifecycle management by professional open innovation intermediaries. The present *Part IV* merges findings from the single cases of *Part III* in a cross-case analysis and contrasts the integrated findings with findings from literature as outlined in *Part II*[592]. This part thereby sets the foundation to derive managerial and research implications in *Part V*.

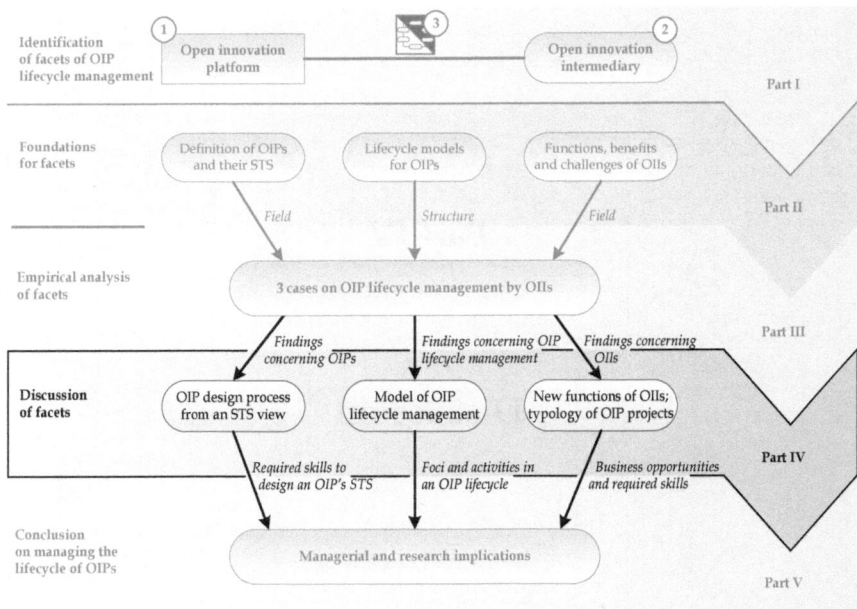

Figure 46: Current progress in the research design

The cross-case analysis is structured along the findings from the single cases in *Part III*. Initially, the focus lies on open innovation intermediaries. The first two chapters reflect on the functions of them (*chapter 1*) and derive a typology of OIP projects as they are run by OIIs (*chapter 2*). The following chapter addresses open innovation platforms and derives a process for designing an OIP from a socio-technical perspective (*chapter 3*). Subsequently, the OIP lifecycle model is set out in order to

---

[592] See Figure 46.

describe activities in each phase of an OIP's lifecycle (*chapter 4*). Finally, mechanisms to overcome major challenges in the lifecycle of OIPs are analyzed (*chapter 5*). *Chapter 6* summarizes the key findings of this part.

The researched open innovation intermediaries fulfill a subset of functions that traditional innovation intermediaries can adopt[593]. This chapter outlines how OIIs fulfill the functions of traditional innovation intermediaries and identifies a new function, namely *providing and managing technology* of OIIs. The key quotes concerning the functions of OIIs, which lay the basis for this chapter, are summarized in Table 27.

Table 27: Key quotes concerning the functions of OIIs

Dimension	Quote concerning the functions of OIIs
Connection	*We want to keep it as diverse as possible. Everybody should be able to participate because, ideas to not necessarily origin from one sector. Often real innovations stem from another sector you did not bear in mind. (A_CEO)*
Collaboration and support	*Furthermore, we organize a workshop to integrate ideas: For example, if you start a project with Atizo, about 500 ideas are gathered. After the client has filtered them about 100 to 150 ideas are taken into the workshop for integrate them. (A_CEO)*
	The workshop is about getting a commitment from specific people and to define who is able to realize a good idea and until which point of time the idea should be realized. All these facts are summarized in a roadmap. We actually wanted the client to create that roadmap, but we noticed that clients sometimes do not have the competences to create such a roadmap. We transfer that capability through trainings. (A_CEO)
	If you look at it from the perspective to get to know the user or that you want creative ideas, you will face a lot of frustration with traditional methods. If I do market research and statistics, I will have nice figures, but after all, it will not represent reality. (I_CEO)
Technological services	*If you want to integrate users into an innovation process in a simple way, without the need to provide your own development team, and without using seven different tools and without having seven different user logins, that is what you get. You will be able to run every user-integration project with us. That is what it shall become. (I_CIO)*
	The client gathered, if at all, experience from a single platform. That is why we say sorry, that is not reasonable and it will not work like this. (H_CPM)

Firstly, in the dimension of *connection*[594], all OIIs provide *gatekeeping and brokering* services, as shown in Table 28. They link organizers that seek solutions to an innovation problem and innovators that can provide a solution to an organizer's

---

[593] See Part II.2.2.
[594] Italic text identifies the quotes addressing the topic as listed in the respective tables and figures.

innovation problem, thus facilitating the flow of knowledge between them on their OIP. OIIs furthermore integrate knowledge from different domains by attracting a vast variety of innovators from different domains and educational levels.

Secondly, open innovation intermediaries *articulate demands* of customers by connecting the organizer with its customers and other external stakeholders. This connective dimension is facilitated by a (typically) open call for ideas and targeted marketing activities.

Table 28: Functions of the researched open innovation intermediaries[595]

Dimension	Function	innosabi	HYVE	Atizo
Connection	Gatekeeping and brokering	✓	✓	✓
	Middle men between science policy and industry	-	-	-
	Demand articulation	✓	✓	✓
Collaboration and support	Knowledge processing and combination	✓*	✓*	✓*
	Commercialization	✓	o	o
	Foresight and diagnosis	-	-	-
	Scanning and information processing	-	-	-
Technological services	Intellectual property	o	o	o
	Testing, validation and training	o	o	o
	Assessment and evaluation	o	o	o
	Accreditation and standards	o	o	o
	Regulation and arbitration	o	o	o
	Providing and managing technology**	✓	✓	✓

* function fulfilled by community
** function extending current frameworks
o complementary function OIIs could fulfill
- function not recommended for OIIs

Thirdly, *knowledge processing and combining* in the dimension of *collaboration and support* is supported by integrating knowledge of the innovators. In most cases, this is accomplished by the innovators and not by the OII itself[596]. Hence, the traditional

---

[595] Building on Bessant and Rush (1995); Howells (2006); Lopez-Vega and Vanhaverbeke (2009).
[596] An exception are community management activities like merging similar ideas that might be done by an OII.

function of innovation intermediaries does not hold anymore: It has shifted from processing and combining knowledge towards facilitating the processing and combination of knowledge by the innovators.

Finally, OIIs can help organizers to *commercialize* their innovation by providing a direct sales and marketing channel on their homepage, as the case of innosabi shows[597]. This represents a shift from the traditional function of consulting on marketing and sales opportunities towards a direct sales activity carried out by the OII.

To sum it up, the OIIs under investigation clearly set the emphasis on *connecting* organizers with innovators. The functions of *collaboration and support* as well as providing *technological services* are not yet exploited. Only innosabi has intentions to support the commercialization of the developed innovations. Functions like *commercialization, scanning and information processing* as well as the *technological services* would complement the current functions of OIIs. Open innovation intermediaries might thus want to adopt these additional functions in order to broaden their portfolio and further serve organizers to build up continuous relationships between them and organizers.

This finding might result from the criteria to select the open innovation intermediaries for this case study[598]. Although this was not a selection criterion, all OIIs that were investigated focus on integrating *consumers* into the innovation process of organizers. Other OIIs, which might for instance focus on integrating *professionals*[599], might fulfill additional functions. This is for instance the case at Innocentive[600], an OII that offers services to transfer intellectual property rights from the innovator to the organizer[601]. Thus, they fulfill at least partly the *intellectual property* function, which calls for further investigation.

The activity of designing and managing an OIP is not explicitly listed in current research on the functions of innovation intermediaries[602]. However, in the field of OIIs, this is a core value proposition. Thus, I propose to add a new function, named *providing and managing technology* in the dimension of technological services, to

---

[597] See unseraller.de/shop/; retrieved November 20, 2012.
[598] See Part III.1.2.
[599] See Part II.1.2.
[600] See Part II.1.3.3 and Part III.1.2.
[601] See www.innocentive.com/for-solvers/intellectual-property; retrieved September 23, 2012 and Lichtenthaler and Ernst (2008).
[602] See Part II.2.2.

the so far defined functions of innovation intermediaries. This function covers all activities to provide and manage technology that facilitate an innovation task. Providing thereby refers to designing and offering technology (e.g. as a software-as-a-service OIP), while managing refers to activities to keep the system running (e.g. providing community management). All OIIs that were investigated fulfill this function by designing and managing OIPs[603]. Subsequently, a typology of OIP projects is derived based on this new function. Benefits and drawbacks of each type of OIP project are briefly discussed in the course of the derivation.

---

[603] See Table 28.

2 Typology of OIP projects

As discussed in the preceding chapter, OIIs fulfill certain functions. The services of all OIIs that were investigated according to those functions are almost identical[604]. However, the approaches of these OIIs differ in major aspects[605]. Thus, these functions do *not* suffice to classify OIIs. As proposed above, a new function *providing and managing technology* is introduced. This function embraces the two major activities in an OIP lifecycle, as derived from the literature[606]: *OIP design* and *OIP management*. The data shows that the OIIs differ in the way they fulfill this function, i.e. the activities within this function[607].

Therefore, I propose a classification of OIIs based on the way they fulfill the function of *providing and managing technology*, i.e. the kind of OIP projects OIIs run. The two dimensions derived from the literature span a matrix to *classify OIP projects*, as depicted in Figure 47: The first dimension concerns the way an OII designs an OIP, namely *OIP design*, the second one the prime responsibility for the *OIP management*. The attributes characterizing the two dimensions are discussed in the following. They stem from empirical findings as Table 29 shows.

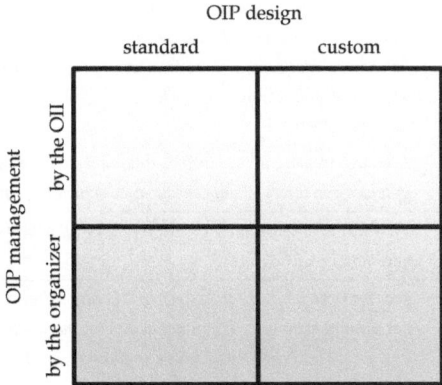

Figure 47: Dimensions and attributes to classify OIP projects

---

The first dimension incorporates the *OIP design*. It can be either a *standard OIP design*, like in the innosabi case, in which all innovation projects run on the same platform, or a *custom OIP design*, like in the HYVE case, in which innovation projects run on distinct platforms, with custom processes and a custom design. Besides those extremes, a customizable standard OIP design constitutes an interim position, like in the Atizo case. A customizable design combines the strength of both extremes.

Table 29: Key quotes concerning the typology of OIP projects

Dimension	Attribute	Quote
OIP design	Standard	*If you want to integrate users into an innovation process in a simple way, without the need to provide your own development team, and without using seven different tools and without having seven different user logins, that is what you get. You will be able to run every user-integration project with us. That is what it shall become. [...] We have a clear picture of what works and what does not. (I_CIO)*
		All of it is already there. Then we have an app-structure: If a company wants a custom idea submission form, we build an app for that. This app overwrites the standard functionality. (A_CTO)
	Custom	*We have many clients that want to start with a small pilot to try the system. The target is to make it a full version after half a year or one year and really customize it at that point. (H_CPM)*
		The internal platforms are our asset. That is why we do all of the development in-house. [...] The degree of customization for internal platforms is way higher [than for external platforms]. There are always processes, rights management and so on. The customer has way more modules to pick from and it is always for the long run. (H_CPM)
		Although we have a huge toolbox full of methods for evaluation, with all sorts of graphics and filters, clients always need something particular. (H_SPM1)
OIP management	By the OII	*We read every contribution of the community, review it and answer every single question. (I_CEO)*
	By the organizer	*We support a lot at the introduction. Having software is one thing, but if you do not life and manage it, it will not work. You need someone [at our client] who manages it actively every day as a community manager. (H_CPM)*
		During the runtime of a project, I check at least once with the client how it works. I contain if they are really active or just wait for the ideas to come in. I advise them to write comments and mark interesting ideas so that the community gets a feedback. (A_ADM)

The second dimension represents the *OIP management*. If an innovation community is implemented on an OIP, the major activity during OIP management is community management[608]. The OIP management can either be done *by the OII*, like in the innosabi case, or *by the organizer*, like in the HYVE case. In addition, interim positions are possible, as the Atizo case shows: The organizer does most of the community management, but the OII also monitors the innovators. Each configuration of the two dimensions yields different benefits and disadvantages for the organizer and the OII as illustrated in Table 30.

Table 30: Benefits and disadvantages of OIP design and OIP management configurations

Dimension	Attribute	Benefits for organizer	for OII	Disadvantages for organizer	for OII
OIP design	Standard	Low effort / Validated OIP design / Short OIP design time / Draw on established community	Scalability / Community building	Not branded / Not adaptable to needs	Reproducible concept
	Custom	Targeted to needs / Branded	Selling advanced services	High effort / Long OIP design time	High effort / Complex projects
OIP management	By the OII	Knowledgeable community management / Low effort	Cross-project community building	Employees' reluctance / Less insights	High effort
	By the organizer	Deep insights into community / Guiding the community	Scalability	High effort / Experience required	Training of the organizer / Monitoring required

Applying those two dimensions and their attributes, the matrix reveals four types of OIP projects, as shown in Figure 48: *standardized, facilitated, self-made* and *custom OIP projects*. Summarizing the benefits and disadvantages listed in Table 30, a *standardized OIP project* yields the benefit of a low effort project for the organizer, which allows scalability for the OII due to a standardized process and OIP. *Facilitated OIP projects*

---

[608] Even if no innovation community is implemented in the OIP's design, support has to be given to innovators of the OIP.

are full service projects to the specifications of an organizer that do not require a particular knowledge in the field from the organizer, as the OII takes most of the work. *Self-made OIP projects* owe their name to the fact that an organizer can run such an OIP project without the help of an OII, for instance by using a SaaS OIP and performing the community management on its own. Finally, *custom OIP projects* require the most knowledgeable organizer, as this is the most individual project type allowing for the most influence and customization by the organizer.

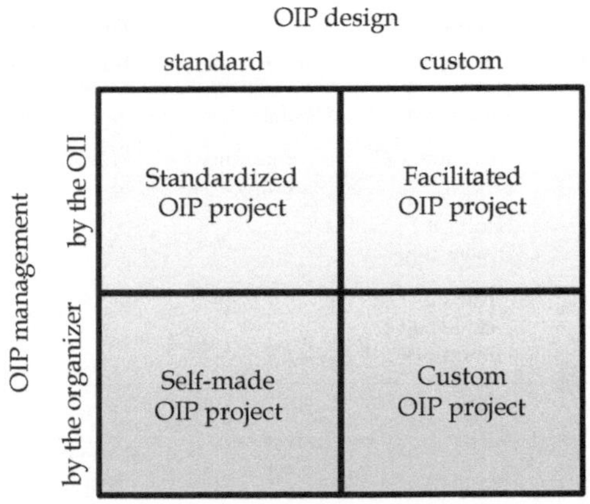

OIP design

	standard	custom
by the OII	Standardized OIP project	Facilitated OIP project
by the organizer	Self-made OIP project	Custom OIP project

OIP management

Figure 48: Typology of OIP projects

Independent of the type of OIP project, designing an OIP should follow a certain process induced by an OIP's socio-technical system. This process of OIP design is introduced next.

3 The OIP design process from a socio-technical perspective

The data shows that an open innovation intermediary has to hold particular competencies in the field of open innovation and open innovation platforms, as exemplified in Table 31. As suggested by Pasmore et al.[609], technical knowledge to design the OIP has to be accompanied by knowledge on the social subsystem of OIPs in order to design and manage it appropriately[610]. The OIIs under investigation build up this expertise with *prior knowledge* in the field of open innovation and OIPs, a strong *integration with research* as well as *iterative improvements* in pilot projects.

Table 31: Key quotes concerning required competencies in the field of open innovation and OIPs

Competency	Quote
Prior knowledge	*The client gathered, if at all, experience from a single platform. That is why we say sorry, that is not reasonable and it will not work like this. (H_CPM)*
	We are on leading edge concerning that topic because we have been in that business for four years and have carried out by far the most projects of them all. (A_CEO)
	It was certainly helpful having technical know-how whereas I doubt the fact whether having this know-how is really necessary. Definitely helpful were experiences from past projects and the experience in open innovation that we collected up to now. (I_CEO)
Integration with research	*We needed some kind of acknowledgement for the users. [...] We also found that in studies: The best would be the company joining a close dialogue with the community, but companies are not up to that. They do not want to explain for 500 ideas why they like the one any why not the other one [...] and that is why we picked the next best reward [...]: money. (A_CTO)*
	That is the typical Atizo process that we have developed in cooperation with Professor Gassmann. Compared to the version Professor Gassmann has documented in his book about crowdsourcing, we use a simplified version. (A_CEO)
Iterative improvements	*That is how unserAller was developed. A lot of prototypes and trial-and-error. Having a spark and then simply trying it. If it did not work, we leave it at that. That is it. (I_CEO)*
	From a development perspective, it [unserAller] evolved step by step with the projects we ran. We had a vision of a platform in mind right from the outset. Nevertheless, we said that the vision is too big for a first step and that we need to gain experience first. (I_CEO)
	We have many clients that want to start with a small pilot to try the system. The target is to make it a full version after half a year or a year and really customize it at that point. (H_CPM)

---

[609] Pasmore et al. (1982). See also Verona et al. (2006) in Part II.2.3.
[610] See Part II.1.4 and Part II.3.6.2.

Building up this knowledge is important since the traditional focus in OIP projects has shifted, namely from the technical subsystem in traditional website projects towards the social subsystem in OIP projects[611]. In OIP projects, the technical subsystem can be viewed as an enabler that reduces barriers of participation for the innovators and enables scalability for an OII[612]. The social subsystem, however, serves as a selection mechanism and motivator for the innovators, which is in turn an important factor to assure the quality of an OIP's output[613].

Table 32: Key quotes concerning the components of an OIP's environment and social subsystem

Component	Quote
Environment	*Then there is the works council. They say if employees submit cool ideas, they have to get incentives. We said that that is not possible, as we want to foster collaboration. If you provide incentives, each employee works on his own credit and no one writes comments on other's ideas. That is why we said we go without. [...] We finally got there and are also backed by the works council by now. (H_CPM)*
People	*We want to keep it as diverse as possible. Everybody should be able to participate because, ideas to not necessarily origin from one sector. Often real innovations stem from another sector you did not bear in mind. (A_CEO)*
Structure	*You can see it at this [screenshot of the] prototype. We have always been rather design driven. A good look was always important. You will have fun to participate and it will somehow work. (I_CEO)*
	What motivates them is fun doing it. If you get such a package and can try out things, it becomes very playful. There is this term of gamification and that is a valid point. If you enjoy doing it, you will do a lot. (I_CEO)
	On the one hand, external ones [OIPs] are completely voluntary. It is for fun only. Internal platforms on the other hand, inherit some kind of pressure from the managers or executives that force employees to use it. Those are two completely different [motivational] approaches. [...] The levers are the same but you have different mechanisms. (H_PM2)
	The challenge is to keep it as simple as possible. That is a key learning from the [project name] project: simpler. Work on the logic that enables the viral mechanisms and then start it. (H_SPM2)

Thus, within the social subsystem, the most important component is the motivational *structure* facilitated by an OIP as already proposed by Verona et al.[614]. Current

---

[611] See the quotes concerning *structure* in Table 32.
[612] See the quotes concerning *technology* in Table 33 and chapter 2.
[613] See in particular the quotes concerning *structure* in Table 32 and *technology* in Table 33.
[614] Verona et al. (2006). See the core competency *creating incentive systems* of OIIs in Part II.2.3.

research highlights the diversity of available options to design the motivational structure of an OIP[615]. Besides this challenge, which consists in designing the motivational structure, a second key challenge is its translating into the *technology*, namely the functions and graphical design of an OIP. The cases show, that there is no blueprint on how to achieve this, but experience is required in order to estimate the impact of manifestations in the technology on the social subsystem[616]. This is evident in the cases that apply a standard OIP (i.e. standard technology), which is continuously improved. OIIs incrementally adjust the technical subsystem, with a special focus on the functions of an OIP in order to evaluate the outcome on the social subsystem. This is a benefit of a standardized approach in running OIP projects: The applied socio-technical system, i.e. the OIP design, is proven, because it was used many times before, and can be further improved in its details[617].

Table 33: Key quotes concerning the components of an OIP's technical subsystem

Component	Quote
Task	*We know pretty exactly how we have to ask in order to come to a result. In most cases our client has already a clear idea in which direction it wants to go. So it comes down to finding the formulation of question that fits the imagination and that we can fulfill with our platform so that the community hast fun participating. (I_CEO)*
Technology	*In the meantime, we have a very standardized scheme how to put up a prototype package and we have a considerably detailed knowledge about which contents have which effects. [...] It has been always important that it looks good, that you enjoy participating. (I_CEO)*
	For instance Pril. What we [would have done] differently, is to implement a jury evaluation into the platform's process in order to neutralize the communities' subjectivity. That would not have happened to us because we understand how communities function, why they do what they do and how they are motivated. (H_SPM1)
	We recognized that we need a form of feedback so that the company can engage with the community. We implemented that, tested it and now it works quite well. (A_CTO)
	I think technology, design and the user interface are important factors to avoid losing a user [...]. Every user I bring to the platform costs money and if he does not get along, or does not find the functionality he wants, or if that functionality is not implemented well, than he will be gone. (A_CTO)

---

[615] See Part II.3.6.2.
[616] See the quotes concerning *technology* in Table 33.
[617] See chapter 2 and the quotes concerning *technology* in Table 33.

To sum it up, an OIP designer, such as an OII, primarily has to design the intended social subsystem of an OIP and design the technical subsystem in accordance with it. Indeed, an OIP designer can only implement the technical subsystem explicitly. The social subsystem is only influenced indirectly by manifestations in the technical subsystem. An OIP lifecycle manager has to be aware of this fact as the major aspects of the social subsystem have to be designed first.

This prioritization of the social subsystem's design *induces* a recommended OIP design process: Certain components of the OIP's socio-technical system depend on another component, i.e. they need to know the design of another component for it to be designed. Consequently, the components that define parameters for others have to be designed before the depending components. The process of designing an OIP from a socio-technical systems perspective is derived from the dependencies of the components. This process, named the *OIP design process*, runs as follows: Initially, the OII has to assess the *task*, such as the problem formulation, with the organizer to reflect the organizer's needs. The task in turn defines the required or intended *people*, i.e. the innovators, who are needed to solve the task. The people required set the premises for an appropriate *structure*, i.e. the motivational system, to motivate those people. Finally, the *technology* implements the structure and thereby reflects the motivational system. Once the OIP's STS is designed in these steps, the OII should check back the fit of the task and the technology and probably adjust both so they match each other. Finally, adjustments considering all components should be made in order to harmonize the overall socio-technical system[618]. Figure 49 depicts the process of designing an OIP from a socio-technical perspective, i.e. the OIP design process.

An *OIP designer* faces the challenge that it can only influence the technical subsystem, namely the technology and task directly[619]. The social subsystem, i.e. people and structures, can only be influenced indirectly. Therefore, an OIP designer might be tempted to focus on the technical subsystem of an OIP and neglect the explicit consideration of the social subsystem.

Current practices at the OIIs under investigation do not explicitly address the need for designing an OIP according to the OIP design process. In the researched OIIs for instance, the task is defined as only one of many aspects of the OIP and not as a

---

[618] See the considerations on the socio-technical systems theory in Part I.3.
[619] See Part II.1.4.

first step[620]. Additionally, organizers are not always explicitly asked what kind of innovators (people) they expect to solve their task[621]. This is defined implicitly by the OII, who guesses the intentions of the organizer.

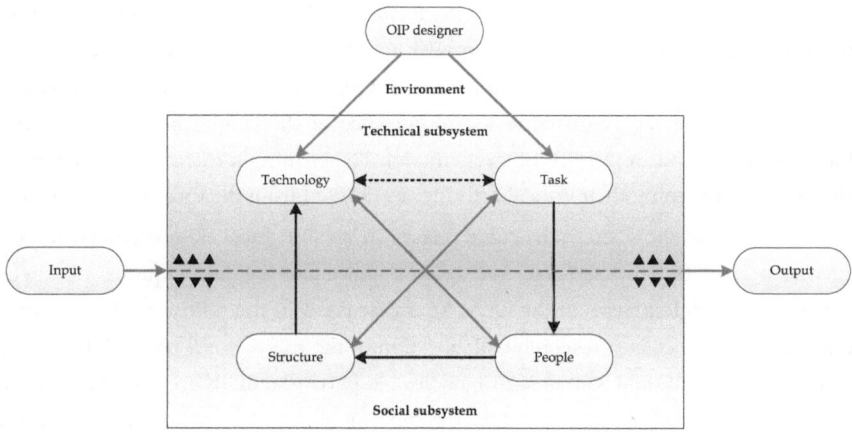

Figure 49: The OIP design process[622]

The findings of the present chapter are particularly relevant for the design phase in the OIP lifecycle model although they span the overall OIP lifecycle, as this thesis in general takes a socio-technical perspective[623]. The subsequent chapter discusses details of each phase in an OIP's lifecycle.

---

[620] See the quotes concerning *task* in Table 33.
[621] See the quotes concerning *people* in Table 32.
[622] Adapted from Bostrom and Heinen (1977).
[623] See Part I.3.

4   The OIP lifecycle model

The aim of this chapter is to set out in detail the OIP lifecycle model so that it can be used by both parties, researchers, who wish to allocate their research in the lifecycle of an OIP, and practitioners, who aim at identifying critical activities in each phase of an OIP's lifecycle. To begin with, the importance of the single phases in an OIP's lifecycle is assessed in the OIP lifecycle model. The ensuing *sections 4.1 to 4.6* explain the phases concerning their critical activities as derived from the three in-depth cases.

The cases show that particular phases in an OIP lifecycle are crucial for the success of an OIP project. Table 34 therefore summarizes the key quotes. The relevance of the single phases of the OIP-LM is discussed in the following. OIIs heavily build on experience and research collaboration in the *requirements phase*[624]. However, compared to traditional web-design agencies, no particular skills are essential here, as default techniques to elicit and manage requirements can be applied. Therefore, the requirements phase is not crucial in an OIP's lifecycle. Although the cases are not definite, the *design phase* appears to be particularly important in the lifecycle of an OIP. This reflects the consideration in chapter 3 according to which designing a meaningful socio-technical system for an OIP is challenging. This is particularly interesting, as current literature provides first insights into OIP design[625]. However, this knowledge is still comparably scarce. Furthermore, a sound architecture facilitates scalability for an OII[626]. Compared to traditional website projects, the *build phase* loses importance, as the design phase gains importance. The quality of the technical implementation can be seen as a hygiene factor. The technical subsystem has to function properly in order to represent the intended social subsystem and not destroy the innovators' motivation[627]. Consequently, an OII cannot establish a competitive advantage by a superior build phase. The activities in the *deploy phase* shift from technical deployment towards marketing activities, which serve to recruit potential innovators. This phase is important, but not as crucial as the design and operate phase. All OIIs in this study unanimously state that the *operate phase* is particularly important and time-consuming in an OIP project. Here, the innovators

---

[624] See section 4.1.
[625] See Part II.3.6.2.
[626] See section 4.1 and 4.2.
[627] See also Table 32 in chapter 3.

have to be motivated to participate in the OIP. Consequently, the literature's view on the importance of facilitating and managing OIPs[628] is warrantable. In standard OIPs, the *optimize phase* is also important, yet, not a crucial phase in the overall process.

Table 34: Key quotes concerning the importance of each phase in the OIP lifecycle model

Phase	Quote
Require-ments	*Building a new platform with having every single task planed is a huge effort, which means that you plan more than you build. Actually, you have to open the source code and write down what comes to your mind, hoping that something appropriate has come out after eight hours of work. (A_CTO)*
	That may be two days or less you have to invest because it has become more of a routine for us. We know pretty exactly how we have to ask in order to come to a result. (A_CEO)
Design	*You can see it at this [screenshot of the] prototype. We have always been rather design driven. A good look was always important. You will have fun to participate and it will somehow work. [...] That is how unserAller was developed. A lot of prototypes and trial-and-error. Having a spark and then simply trying it. If it did not work, we leave it at that. That is it. (I_CEO)*
	We now know what the product [unserAller] will look like and have the vision how it shall be and will be. We did not have that for the first version. [...] At the beginning of the year, when large clients approached us, scalability became a subject [...] and that is why a complete redesign from scratch was justifiable anyway. (I_CTO)
Build	*That means that tasks like programming do not have to be done exclusively in-house, as it was in the old days. It was relevant when you have had a benefit through having the know-how which came along with programming. Nowadays, the actual know-how is the process itself, the consulting and everything around it while code per se is exchangeable. (H_SPM1)*
Deploy	*One client said that his CEO wrote an e-mail to all employees and pushed the topic. The other one said that he walks from desk to desk and tells each employee the whole story and helps them to submit their first idea. Those are two completely different approaches, which both led to success. (H_CPM)*
	We were able to build the community really fast without spending a single Euro for advertising. This was actually supported by the viral effect on Facebook. (I_CEO)
Operate	*Once the contest goes live, the work really starts. (H_CPM)*
	We read every contribution of the community, review it and answer every single question. If we have 450 design proposals and 2,500 comments like in the [name of project] project, we have to approve all of them. This is quite a big effort [...]. (I_CEO)
Opti-mize	*We have frequent releases once the [internal] contest is live because the customer says we now need this and that feature. Then we implement it accordingly. (H_CPM)*
	There are always those little bugs and changes, [...] but we can usually solve and deploy them within minutes. (I_CTO)

---

[628] See Part II.2.3 and Part II.3.6.5.

Figure 50 depicts the OIP lifecycle model including the prioritization of lifecycle phases as discussed above. The phases in the OIP lifecycle model do not necessarily run consecutively, but can overlap due to parallel circles (e.g. multiple changes are implemented at the same time) or iterations (e.g. another circle starts before the prior one has finished or two or more process steps have to be iterated)[629].

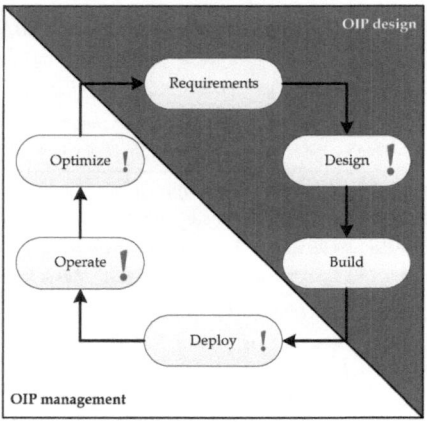

Figure 50: Prioritization of phases in the OIP lifecycle model

At this point, it becomes clear that the skills required to manage the lifecycle of OIPs differ from those to manage the lifecycle of a default website. While default web-design agencies focus on the *design and build phase*, OIIs should focus on the *design and operate phase*. Furthermore, they should make effort in the *phases deploy and optimize*. The following sections discuss details of each phase in an OIP's lifecycle.

4.1   Requirements phase of OIPs

When running an OIP project, *non-functional* and *functional requirements* have to be elicited, as evident in Table 35 and Table 37. Thereby, the focus lies on functional requirements and the OIP's social subsystem, as outlined in chapter 3.

The most important *non-functional requirements* for OIPs are *usability*, with a focus on simplicity and intuitive handling, and *accessibility*. These core non-functional requirements should be targeted in all OIP projects. The ultimate aim should be to attract the right innovators and make them enjoy innovating.

---

[629] See Part II.3.4.4.

Table 35: Key quotes concerning non-functional requirements

Requirements	Quote
Non-functional requirements	*Technology, design and the user interface are important factors to not lose a user. You cannot be good enough in them. Every user I bring to the platform costs money. If he does not get along or does not find functionalities he wants or the functionalities are not implemented well, I will lose him. (I_CIO)*
	You can see it at this [screenshot of the] prototype. We have always been rather design driven. A good look was always important. You will have fun to participate and it will somehow work. (I_CEO)
	From a technical point of view Facebook is a disaster: APIs are changed without reasons and with a 2-month notice. The developers have to rebuild parts of the platform within that timeframe. When certain functionalities have changed or are not available any more, you have to be flexible and to react on short notice, whereas this additional effort is not necessary. In addition to that, you need to do constant bug tracking as there are random problems [with the API of Facebook] you have to solve for your clients. (I_CEO)
	For example rivella.atizo.com or the milk serum drink from Switzerland are easy to set up because you set the branding, the logo, the colors and the background while the actual configurator is already ready to use. (A_CTO)
	Putting different themes on a working technical base is something you need extraordinarily often in user-integration projects. (I_CTO)

In standard OIPs[630] or OIPs that build on a common framework, *maintainability* and *modularity* are crucial to allow for scalability of the OII and a continuous optimization of the OIP. Table 36 outlines the most important non-functional requirements that an OIP's architecture should implement.

Table 36: Non-functional requirements of OIP's

Non-functional requirements	
Usability*	Navigability
Accessibility*	Security
Modularity*	Confidentiality
Maintainability*	Scalability
	Extendibility

* important

Functional requirements have to be assessed in particular for the *frontend*, the *backend* and the *integration with an organizer's IT and processes*[631]. The functional requirements

---

[630] See chapter 2.
[631] See Table 37.

(i.e. the technology[632]) depend on the aim of the OIP, the supported phases of the innovation process and the OIP's intended social subsystem. In order to cover all important functional requirements for the *frontend*, requirements elicitation should follow the OIP design process as described in chapter 3. Therefore, the aim (*task*) of the OIP should be clarified at first. Subsequently, the required innovators (*people*) have to be defined. Thereafter, the motivational *structure* should be assessed for the functional requirements (*technology*) to be designed in the end. The requirements can remain on a relatively abstract level, as some creative freedom is required to design the OIP in a harmonic way. Details of the functions, which are part of the technical subsystem, are determined once the OIP is in the design stage. Accordingly, the overall process of OIP design should follow an agile approach[633].

Table 37: Key quotes concerning functional requirements

Requirements	Quote
Frontend	*We have two types of clients. The one type approaches us and wants us do build an inno-vation community platform. This is a typical me-too, as others have it as well. [...] The other type of client approaches us and says: this is our process and this is what the tool should look like. They have some 50 pages requirements document because they invested months in thinking it through. (H_CPM)*
	[The client] tells us which functionalities he wants and how he imagines the project. Then we build it the way that it is perfect for him and hope that it is transferrable to our other clients. (I_CEO)
Backend	*Although we have a huge toolbox full of methods for evaluation, with all sorts of graphics and filters, clients always need something particular. (H_SPM1)*
Integration into an organizer's IT and processes	*The degree of customization for internal platforms is way higher [than for external plat-forms]. There are always processes, rights management and so on. The customer has way more modules to pick from and it is always for the long run. (H_CPM)*
	An unserAller project involves different departments which ex ante define their solution space and which also take the final decision in each phase. [...] It is a relatively complex process. (I_CIO)
	Other clients say that it is about ideas, which are their core asset. It has to be hosted internally. Especially when we integrate single-sign-on. (H_SPM2)

Besides eliciting requirements for the frontend, requirements for the *backend* have to be assessed. A supportive administration backend is crucial for the operations phase (especially the community management) and thus for the success of an OIP project.

---

[632] See chapter 3.
[633] See Part II.3.3.

In addition to that, requirements for the *integration into an organizer's IT infrastructure and processes* have to be clarified. The process of requirements elicitation and management can be standardized. All three OIIs that were investigated apply different approaches although each one is standardized in its own way. Table 38 outlines the major questions that should be addressed in the requirements phase.

Table 38: Major questions to clarify during the requirements phase

Area	Question
Frontend	What is the task?
	What are the required characteristics of the innovators?
	What is the required motivational structure? Are the innovators driven by non-monetary or monetary rewards? What kind of experience do they want to make?
	What are the intentions of the organizer? Should the OIP be branded or not?
	Which phases of the innovation process should be supported?
	Should the OIP bridge the gap between the online and offline world? How?
	What are the organizer's corporate guidelines/ corporate identity requirements?
	What are the legal constraints?
Backend	What kind of community management should be done? Shall all contributions be reviewed? What kind of feedback should the community get?
	What is the role of the community manager? Active? Passive? What are the community manager's activities?
	Who performs the community management?
	Which functionality is required to manage the community?
	Which reports and analyses are required during and at the end of a project?
	How to identify and deal with fraud?
Integration into an organizer's IT and processes	What are the dependencies with the organizer's IT? Is single sign on possible?
	How to make sure that the communities' results are adopted by the organizer?
	Who needs to be integrated in the OIP project? Who is required to manage the community?
	Who is required for processing of the results/ implementation of the community ideas into the organizer's processes?

In order to clarify the questions, different default methods of requirements elicitation are applicable. Due to the ambiguous character of the social subsystem, OIIs should integrate *feedback from potential users* of an OIP in order to elicit requirements and gather feedback as shown in Table 39. Pure bulletin boards that allow innovators to articulate feature requests do not suffice for this. On the contrary, to gain substantial input, serious integration of and personal contact with innovators and organizers is indispensable.

Table 39: Key quotes concerning the source of requirements

Source	Quote
Feedback from innovators and organizers	*We have a list of 15 beta testers. Those are mainly companies, which are completely different. We have software developers who want to test the interfaces, caterers, small producers, large producers and also community members. (I_CEO)*
	We have frequent telephone calls with our top users. They are enthusiastic to tell us what they want to change and which new features they want to have. But they do not do that via a standardized platform. (I_CEO)
	What we also do is a biannual user meeting. We invite all customers that are interested. We bring them together to exchange experiences. On the one hand, this exchange is very valuable for the client; on the other hand, we [HIC] get an impression of the bugs and real problems. (H_CPM)
Results of research	*We needed some kind of acknowledgement for the users. […] We also found that in studies: The best would be the company joining a close dialogue with the community, but companies are not up to that. They do not want to explain for 500 ideas why they like the one any why not the other one […] and that is why we picked the next best reward, which is divisible and internationally distributable: money. (A_CTO)*
	[…] in the meantime [we do] not [track the OIP] that exactly anymore because we currently do not completely analyze the data. We partly analyze data for research projects that we carry out and run intensive website tracking therefore, but in principle it is easier to ask our users directly to explore problems because they know pretty exactly why something does not work. (A_CTO)
	That is the typical Atizo process that we have developed in cooperation with Professor Gassmann. Compared to the version Professor Gassmann has documented in his book about crowdsourcing, we use a simplified version. (A_CEO)

In addition to the traditional requirements elicitation methods, important sources of information are *results of research* concerning open innovation[634]. The literature discusses many aspects for the design of an OIP and motivational mechanisms. Therefore, an OII should have a strong link to research in the field. Besides using results of research, the OIIs under investigation actively contribute to the research community. This might be a strategic decision to gain access to otherwise unavailable resources. However, knowledge on the backend of an OIP has not been investigated yet and needs to origin from other sources or experience.

---

[634] See Table 39.

Over all, as being common in agile software development approaches[635], OIIs should not overemphasize the requirements phase. They should rather create early mockups or build a prototype to discuss a design with an organizer and improve it incrementally. Thereby, an OII should guide the organizer and try to make the OIP as lean as possible. Table 38 summarizes the key aspects during the requirements phase. The following phase in the OIP lifecycle model is the design phase, which is discussed below.

Table 40: Major activities in the requirements phase

Activity in the requirements phase
Follow the OIP design process
Draw on research
Focus on functional requirements
Major non-function requirements are usability, accessibility, maintainability and modularity
Elicit requirements for frontend, backend and integration into an organizer's IT and processes
Do not overemphasize requirements

4.2   Design phase of OIPs

The assumption that professional OIIs *rely on trial-and-error* approaches and their experience[636] has been confirmed[637]. There is a vast amount of options to design an OIP with no clear guidelines on what to do. Research provides some insights into meaningful configurations, but mainly in the shape of concepts which can only serve as rough ideas. Details are still subject to an individual design. The challenge is to meet the requirements of a social subsystem by designing a technical subsystem. An *iterative design*, applying a perpetual-beta paradigm, should be applied to design the user interfaces and motivational structures of an OIP. This is due to the ambiguous character of design decisions, the absence of clear guidelines, and the only indirect impact of the technical subsystem on the social subsystem[638]. The OIP's architecture in turn should support this volatile environment.

---

[635] See Part II.3.4.
[636] See Part II.3.2.
[637] See Table 41.
[638] See Part II.1.4.

Table 41: Key quotes concerning the design phase

Aspect	Quote
Relying on trial-and-error	*That is how unserAller was developed. A lot of prototypes and trial-and-error. Having a spark and then simply trying it. If it did not work, we leave it at that. That is it. (I_CEO)*
	We are currently working on a project to automatically group short texts. We have first promising results that we will integrate in the next months to see if that is really useful. (A_CTO)
Iterative design	*I cannot exactly tell how long [the designer] needs for a design. But I know that none of the designs that he delivers will be the same in the next week. Once it is live, we act very fast. (I_CTO)*
	We try to have a first version, like a prototype, as soon as possible to publish it to the client. You can think about a lot of things in theory but the client has to see it in order to exactly tell what he wants. (H_CPM)
Architecture	*We now know what the product [unserAller] will look like and have the vision how it shall be and will be. We did not have that for the first version. [...] At the beginning of the year, when large clients approached us, scalability became a subject [...] and that is why a complete redesign from scratch was justifiable anyway. (I_CTO)*
	We actually offer it as software-as-a-service. In principle, we could do it [the implementation] on-site, but so far, we convinced all of our clients that we do not have to do that. (A_CTO)
	All of it is already there. Then we have an app-structure: If a company wants a custom idea submission form, we build an app for that. This app overwrites the standard functionality. (A_CTO)
Channel to innovators	*We recognized that we need a form of feedback so that the company can engage with the community. We implemented that, tested it and now it works quite well. (A_CTO)*
	Although there is no contact between the company and community at great length [referring to commenting every single idea], we try to keep the contact as close as possible in order to express the appreciation. (A_CEO)
	The basic idea of unserAller arose from the observation that people identified themselves with co-created products. We supposed that it would work even better for convenience goods. Meaning things, I build a relation to and with which I identify myself and my lifestyle somehow. (I_CEO)
Type of OIP	*On the one hand, external ones [OIPs] are completely voluntary. It is for fun only. Internal platforms on the other hand, inherit some kind of pressure from the managers or executives that force employees to use it. Those are two completely different [motivational] approaches. [...] The levers are the same but you have different mechanisms. (H_PM2)*
	The internal platforms are our asset. That is why we do all of the development in-house. [...] The degree of customization for internal platforms is way higher [than for external platforms]. There are always processes, rights management and so on. The customer has way more modules to pick from and it is always for the long run. (H_CPM)

In order to facilitate reusability and a common technological basis, the OIP's *architecture* should follow software design patterns. While modularity and a framework-based OIP architecture[639] allow for a scalable approach, multitenancy has shown to be the silver bullet to design an OIP, as the Atizo case shows: The OIP is adaptable to an organizer's needs while the technological basis remains common. Furthermore, improvements that have been achieved for one organizer can be used for all organizers. The non-functional requirements listed in Table 36[640] facilitate these design aims.

From a theoretical perspective, an organizer has to be aware of the decision between a direct and a mediated *channel to innovators*[641] when designing an OIP. A mediated channel yields the benefit of a perceived neutrality of the innovation project (like in the Atizo case), while a direct channel with a custom (branded) OIP yields the benefit of marketing and the attraction of innovators by an organizer's brand (like in the HYVE case). In addition, an interim position is possible, as the innosabi chase shows: Although the OII provides the perceived neutrality of an intermediary, it tends to create a direct channel of the organizer to the organizer's customers.

The data confirms that the *type of OIP* matters when designing an OIP, as proposed by prior research[642]. Thus, a default approach to design an OIP is not applicable. The OIIs under investigation distinguish two types of OIPs. Table 42 summarizes major aspects that might have to be considered depending on the type of OIP: An aspect can be relevant for *all OIPs* or only for OIPs that are to integrate particularly external innovators[643] (*external OIP*) or to those that are supposed to integrate an organizer's employees, i.e. in particular peripheral and internal innovators (*internal OIP*). Additional design decisions, as for instance described in literature on OIPs, still apply[644].

---

[639] See section 4.1.
[640] See section 4.1.
[641] See Part II.2.3.
[642] See Part II.3.6.2.
[643] For a distinction of the types of innovators see Part II.1.2.
[644] See Part II.1 and Part II.3.6.

Table 42: Major aspects to consider in an OIP's design depending on the type of OIP[645]

Type of OIP	Aspect to consider in design	
All OIPs	Frontend	Task definition
		Selection of tools (innovation contest, innovation community, …)
		Timing of phases
		Channels to submit contributions
		Graphical design (playful experience, serious design, …)
		Distribution of responsibilities in community management
		Combination of online and offline components
	Backend	Reports
		Statistics
		Documentation of results
		Review of contributions (none, manual, semi-automatic, automatic)
		Marketing (newsletters, personal messages, …)
		Managing users (modifying, deleting, …)
External OIP	All	Visibility of organizer for the innovators
		Branded vs. default graphical design
		Alignment with an organizer's corporate identity
	For general audience	Motivational system
		Graphical design
	For professionals	Intellectual property
		Professional services (handling technology transfer, …)
Internal OIP	All	Integration into organizer's processes, especially the innovation process, leadership processes and governance
		Integration into organizer's IT (single-sign-on, …)
		Security
		Alignment with work council
		Intellectual property

The designer of an OIP has to think out of the box in order to attract relevant innovators and to creatively support different phases of the innovation process[646], in

---

[645] The design of the OIPs is derived from the OIPs and their respective backends, namely gemeinsamselten.de, unseraller.de and atizo.com. In addition, quotes from the interviews build the basis for this table, like for instance those listed in section 4.1.

[646] See Part II.1.2.

particular the implementation phase. One approach to foster implementation for instance is to connect the online and offline world. The cases show different strategies to do so, also on a large scale. By adding offline components, like for instance prototype packages in the innosabi case, OIPs gain further importance in the field of open innovation, as they can integrate activities that were typically attributed to offline open innovation methods, like the lead user method[647]. By integrating offline activities, for example, transferring tacit knowledge becomes possible[648]. Bringing these to the online world, combines strengths of both environments.

All OIIs under investigation implement innovation communities in their design. However, the extent to which they draw on this tool for IT-based open innovation differs[649]. The innovators can collaborate to develop a single innovation, like in the innosabi case. However, the innovators can also submit multiple suggestions in a contest setting, like in the Atizo and HYVE case. Thus, a more competitive approach comes into being[650]. These examples show that the implementation of the tools for IT-based open innovation[651] might vary. The designer of an OIP can freely combine IT-based tools open innovation and adjust the intensity of usage concerning each tool. By doing so, the OIP combines the strength of different tools. Table 43 summarizes the key aspects during the design phase. Having fixed a design, the build phase starts as discussed in the following.

Table 43: Major activities in the design phase

Activity in the design phase

Select and combine tools for IT-based open innovation

Perpetual-beta paradigm with early prototypes

Design for simplicity and usability

Strive for an extendable, modular architecture

Draw on research

Design the social subsystem prior to the technical subsystem

---

[647] Luethje and Herstatt (2004).

[648] See Part II.2.4.

[649] This is derived from the OIPs themselves.

[650] This was already proposed by prior research as outlined in Part II.1.3.2.

[651] See Part II.1.3.

4.3   Build phase of OIPs

As theorized in Part II.3.3, the researched OIIs apply an *agile development* approach. They use Scrum, or Scrum-like methods, with only very little documentation, as shown in Table 44.

Table 44: Key quotes concerning the build phase

Aspect	Quote
Expertise in open innovation	*First of all, I brief the designer to create mockups. […] In a second step, we do a briefing where I explain as many points from the concept as possible to the developer, who then develops the platform. (H_PM2)*
	The developer is only responsible for enough requirements in his topic. We implement it all together, dedicating us to a single topic at a time. (A_CTO)
Agile development	*It was an extremely quick development and we did not know what the outcome will be. […] That is why we had to make a great deal of quick changes to the system. That called for plain PHP programming. (I_CIO)*
	We have tasks and stories. We estimate them and have a roadmap to plan our development. That is how we coordinate us. (A_CTO)
	We have frequent releases once the [internal] contest is live because the customer says we now need this and that feature. Then we implement it accordingly. (H_CPM)
Outsourcing	*We actually have a high technological competency in our team. We can quickly realize many things by ourselves and iterate them and try out things and see what works best. That is possible because we do not always have to ask an agency to program it for us. (I_CEO)*
	The internal platforms are our asset. That is why we do all of the development in-house. (H_CPM)
Testing	*We mostly identify bigger bugs internally if we give it [the new functionality] to [a non-technical employee]. He just needs five clicks to find something we have never seen before. He is our best bug reporter. (I_CTO)*
	We have a list of 15 beta testers. Those are mainly companies, which are completely different. We have software developers who want to test the interfaces, caterers, small producers, large producers and also community members. (I_CEO)
	We have a very detailed coverage of automated tests. […] There are not many bugs that go live. If we have a bug in the live system, it is mostly very hidden, for instance in an administrative function, actually things almost nobody uses. (A_CTO)

This is fine since an OIP is a rather simple software from a technological point of view[652]. Notwithstanding, the developers are, and should be, experienced because they work in small teams and cover broad responsibilities. This is in accordance with

---

[652] See the definition of an organic mode project in Part II.3.3.

the implications on project typologies of Boehm and Turner[653]. Compared to tra-ditional web development projects, the agile setting does not constitute a difference.

However, developers in OIP projects should additionally have *expertise in open innovation*. The data show that results achieved by a knowledgeable developer (e.g. the developer and (co-)designer of the OIP are the same person) are more valuable than those achieved by an OIP designer advises a developer, for instance with mockups.

OIP projects call for many small releases in an *agile development* approach to test and improve the socio-technical system. Therefore, agility is the key. This kind of agility, with potentially multiple releases per day, requires support from a corresponding infrastructure. Consequently, all OIIs use a repository in order to manage their code. Furthermore, OIIs can draw on the potential of a comprehensive configuration management system to realize the benefits of multitenancy discussed in section 4.2.

Outsourcing of development tasks is not an option for the OIIs under investigation. The agility discussed above and the expertise that is needed hamper it. In addition, the key competence which is the technology itself, but also knowledge about the technology, should stay in-house. However, OIIs draw on open source components, which they reuse in their architecture.

Finally, *testing* must not be neglected. As outlined in Part II.1.2, OIPs depend on the voluntary contributions of innovators. The barriers of participation should accordingly be as low as possible. Each bug constitutes a barrier that should not exist. In the build phase, testing is important in order to release as little bugs as possible. OIIs can identify many bugs by internal testing. Additionally, automated tests improve the bug spot rate further and are applicable for OIPs, as the Atizo case shows. Consequently, the OIIs release only few minor bugs. OIIs should learn from this example not to pressurize organizers too much into testing the technology of the OIP in the deploy phase. An even worse case could be the loss of an innovator in an advanced phase[654]. Table 45 summarizes the key aspects during the build phase. When the OIP is tested and ready, the deploy phase starts, as discussed next.

---

[653] Boehm and Turner (2003). See Part II.3.3.
[654] See Table 35 in Part IV.4.1.

Table 45: Major activities in the build phase

Activity in the build phase
Engage experienced and knowledgeable developers
Use an agile software development approach
Keep development work in-house
Establish comprehensive testing

4.4 Deploy phase of OIPs

From a technical point of view, *deployment* is rather easy for external OIPs, as dependencies with other systems are rare. Consequently, committing new functionality in a repository suffices for deployment. Contrarily internal OIPs potentially call for more effort, as they might have to be integrated into an organizer's infrastructure (e.g. active directory) or processes (e.g. an innovation process). In both cases, the employment of a software-as-a-service architecture is recommendable in order to ease deployment in a setting of a perpetual-beta paradigm with potentially many releases[655]. Table 46 outlines quotes regarding this aspect of the deploy phase and the ones below.

The emphasis of activities in the deploy phase differs significantly from that in traditional web projects. The utmost concern in the deploy phase of OIP projects is the *attraction of innovators* who are to contribute to the OIP[656]. The different OIIs apply different methods to attract innovators, though the basic strategy remains the same. Firstly, newsletters that announce the new innovation project are sent out on existing communication channels on a large scale. In addition, for external OIPs, a social media campaign is performed on existing social media channels, such as brand communities. Besides that, the intention of the OIIs is to create a viral effect to reach innovators that are somehow related to addressees of the marketing campaign.

An additional, promising approach supports the attraction of innovators: An OII can identify communities consisting of individuals who are potentially relevant to the task and perform targeted marketing within those communities. By that, an OII can actively attract innovators from different domains. Thus, a purposeful and conscious composition of innovators becomes possible, which is for instance an advantage of the lead user method. The case of HYVE shows that the approach,

---

[655] See section 4.3.
[656] This finding is in line with the claims of researchers outlined in Part II.2.3 and Part II.3.6.4.

which identifies relevant communities for targeted marketing, combines the benefits of self-selection mechanisms and purposeful selection of innovators[657].

Table 46: Key quotes concerning the deploy phase

Aspect	Quote
Deployment	*We actually offer it as software-as-a-service. In principle, we could do it [the implementation] on-site, but so far, we convinced all of our clients that we do not have to do that. (A_CTO)*
	There are always those little bugs and changes, [...] but we can usually solve and deploy them within minutes. (I_CTO)
Attraction of innovators	*We were able to build the community really fast without spending a single Euro for advertising. This was actually supported by the viral effect on Facebook. (I_CEO)*
	One client said that his CEO wrote an e-mail to all employees and pushed the topic. The other one said that he walks from desk to desk and tells each employee the whole story and helps them to submit their first idea. Those are two completely different approaches which both led to success. (H_CPM)
	Rivella is a good example for that. They wanted to build their own community from their Facebook fans. They started with a public project and only then users registered with their custom platform. [...] Now users are active on both, Atizo and Rivella's platform. (A_ADM)
	[The client] asked us to create a video clip for the internal communication of the project. (I_CEO)
Training of community managers	*We educate the idea managers on the platform, so they can message the users every day. They evaluate ideas, promise incentives if so and so many ideas are submitted and so on. (H_DEV)*
	If there are international requests or if the client does not have any budget, we do not run any workshops. Therefore, we have an online section called "instructions". There are videos and fact-sheets for our clients. (A_CEO)
Kick-off	*The CEO wrote a letter. That kicked-off the platform and they had 6.000 users on the first day. Subsequently, the middle management pushed the employees to participate, as it would help the overall company. That is a benefit, as the CEO is too far away from most employees. What really counts is that what your line manager tells. (H_CPM)*
	It was a good thing that we were able to start the platform with running projects. Otherwise, we would not have succeeded. (I_CEO)
	For the start of the project, we organized a launch party on the first of June 2010 and invited our whole network. It was like: OK, let's get it started – that was the initial kick-off. (I_CEO)

The attraction of relevant innovators is particularly important, as the quality of the OIP's output is also influenced by the quality and diversity of the input, e.g. the knowledge, characteristics and skills of the innovators. For internal OIPs, support of

---

[657] See Part II.3.6.4.

top management, e.g. by writing newsletters, is required to attract relevant innovators as well as direct line managers should personally advise their subordinates to contribute to the OIP. From a timing perspective, community-building can start prior to the technical deployment of the OIP. A temporary homepage might be applied to announce an upcoming innovation project. To sum it up, a sound marketing strategy to attract relevant innovators has to be addressed explicitly.

Besides activities that attract relevant innovators, *training of community managers* has to be performed. The community managers need an introduction into their tasks, responsibilities and communication rules. The OIIs under investigation make this their personal task. Written documentation on core aspects in community management might complement the personal training.

Once the runtime of an OIP starts, some content should be provided prior to the *kick-off* so that first users of the OIP see some sample content and not have to be the ones who make first contributions. This lowers barriers of participation. The content might origin from the OIP project team, from a group of beta testers or from the organizer's or OII's employees. All innovation projects investigated are launched on a fixed and communicated date. This serves the purpose to create a strong initial kick-off including, for instance, an intensive discussion. Table 47 summarizes key aspects to consider in the deploy phase of an OIP. The next phase in the OIP lifecycle model, namely operate, potentially calls for the highest effort in an OIP's lifecycle.

Table 47: Major activities in the deploy phase

Activity in the deploy phase
Integrate OIP into organizer's IT and processes
Perform multi-channel marketing
Train community managers
Pre-occupy content
Launch on a fixed date

4.5 Operate phase of OIPs

The operate phase of OIPs is more challenging than the one in default web projects, as a critical new task, i.e. the *community management*[658], has to be accomplished[659]. This and further aspects are illustrated in Table 48. The effort of community management makes the operation phase one of the most time consuming phases in an OIP lifecycle. Community management is performed by a new role in an OIP lifecycle, namely the community manager. Community managers actively motivate innovators on an OIP, provide support if content or administrative questions arise and monitor the community. They thereby facilitate the generation of results that are valuable for the organizer.

Different mechanisms can be applied to achieve the goal of valuable contributions for an organizer. Community management can be the organizer's, the OII's or a shared responsibility, as discussed in chapter 2. Innovators might support community management. A community manager can actively direct the innovators towards a desired output by *monitoring and intervening*. This can be accomplished either by commenting on contributions, by evaluating contributions or by filtering desired or undesired contributions in a review process. If an organizer itself manages the community, the organizer knows best which contributions technically and strategically fit its organization. Therefore, an alignment of a community manager with the organizer is not necessary. If an OII takes over the community management for the organizer, it can compensate the missing alignment by frequently integrating the organizer in review loops, as the innosabi case shows. All mentioned approaches seem to improve the implementation rate of an OIP's output at the organizer[660]. In all settings, the OII or the organizer has the possibility to overrule decisions by the innovators, without offending them.

---

[658] Community management does not only apply to OIPs that include an innovation community, but to all OIPs, as questions of innovators have to be answered and support has to be given in any case.

[659] This finding is in line with Verona et al. (2006), who claim that managing two-way communication is important for open innovation intermediaries as outlined in Part II.2.3.

[660] For a more detailed discussion, see section 5.3.

Table 48: Key quotes concerning the operate phase

Aspect	Quote
Community management	*Once the contest goes live, the work really starts. (H_CPM)*
	We support a lot at the introduction. Having software is one thing, but if you do not life and manage it, it will not work. You need someone who manages it actively every day as a community manager. (H_CPM)
	[…] like for instance, a producer of yoghurt we talked to: he can process nuts on his machine, but he cannot process popcorn. No user can know this. (I_CEO)
Monitoring and intervening	*We had this suggestion in our snack project. A snack for 'after having sex'. It was of course very popular and many users clicked on it because they thought it is cool. However, this is no need of the consumers. You can see it from the comments. That is why we deleted it. […] and this is accepted by the community. (I_CEO)*
	We read every contribution of the community, review it and answer every single question. If we have 450 design proposals and 2,500 comments like in the [name of project] project, we have to approve all of them. This is quite a big effort, but that guarantees us that we can quickly react on questions, that we can avoid disputes and that the community feels taken seriously. […] but it also deals with remaining a feeling for the community. Thereby, we can realize if something gets out of control. (I_CEO)
	We cannot monitor everything, but if the community or the moderator reports something, we will have a look if our general terms and conditions are violated. […] The community governs itself very well. (A_ADM)
Motivate innovators	*You have to write a newsletter to the community every week. Hey hello, this is the innovation community. This is our status. Thanks for being with us. Those who are not in yet, please participate. Those who participate, please submit your ideas. Those who submitted ideas, please evaluate ideas. (H_PM2)*
	At its core, [the offline community meeting] is merely a getting in touch, an exchange among birds of a feather. It is just good to meet and talk offline instead of writing online all the time. (A_ADM)
Administration backend	*If we develop new functionality, at the same time we will create a corresponding backend interface. Otherwise, we would not be able to handle the user support if they do not find something. (I_CIO)*
Intellectual property rights	*Yes, [the intellectual property rules] are in the AGB whereas they actually do not matter because you never publish complete ideas, suggestions or concepts but single thoughts and comments. The text that is uploaded by the users is in most cases not longer than three lines. Thus, intellectual property rights do not even constitute in 99 percent of all cases. (I_CEO)*
	If the contribution to a discussion is indeed more than a simple suggestion, the company modifies the suggestion and combines this contribution with other ideas until a final version comes out that merely expresses the users' wishes. Our client never uses the original user's suggestion. (I_CEO)

By constant *monitoring and intervening*, community mangers rule out problems, which would hinder a constructive community culture[661]. However, philosophies on how to conduct community management differ across OIIs. While an OII can review all contributions, as in the innosabi case, other OIIs might rather draw on the self-regulation of a community, as in the Atizo case. Fraud is not considered a problem in the search and implementation phase. However, during a selection phase, measures have to be adopted to prevent manipulation of evaluation results. An additional benefit of constant monitoring by an OII is the OII's ability to recognize and actively gather feedback for further optimizations of the OIP. This feedback serves as input for the optimization phase.

Furthermore, a community manager has to *motivate innovators*. This can be done by newsletters, individual messages or personal communication. OIIs can apply semi-standardized messages to motivate innovators of an OIP. However, completely standardized messages do not yield the intended results, as all cases show.

To facilitate a community manager's tasks, a community manager needs administrative functionality (like statistics, user administration, support of review tasks, export of ideas, etc.), which might be provided in an *administration backend*. Eventually when it comes to the selection of winning contributions, reports are required to document and discuss results.

Though research stresses the importance of *intellectual property rights* and their active management[662], the data do not suggest any need for an emphasis on this topic. OIIs ensure that intellectual property rights are transferred to the organizer. In most cases, contributions of single innovators are so marginal that intellectual property rights do not even constitute. Thus, research and practice diverge, which calls for further investigations. One reason for this finding might be due to the type of OIP. The OIIs under investigation predominantly integrate hobbyists[663] with their OIPs, who might not be striving for a commercial exploitation of their ideas. This might differ for OIPs for professionals. Thus, a contingency approach concerning the type of OIP might be required in the intellectual property rights discussion.

In order to facilitate the implementation of an OIP's output, mechanisms to overcome employees' reluctance, as discussed in more detail in section 5.3, have to be

---

[661] See Part II.1.3.2.
[662] See Part II.2.2 and Part II.3.6.4.
[663] Hallerstede et al. (2010).

implemented. In addition, an OII has to develop competencies in analyzing generated content in order to recombine and transfer it to the organizer[664]. Table 49 summarizes key aspects that have to be taken into consideration during the runtime of an OIP. During the runtime, aspects for optimization have to be gathered, as discussed in the following.

Table 49: Major activities in the operation phase

Activity in the operation phase
Perform active community management
Motivate innovators
Consider intellectual property rights
Gather feedback for improvements
Use a dedicated administration backend
Facilitate implementation of results

4.6   Optimize phase of OIPs

OIPs are subject to a perpetual beta paradigm that requires many small releases for *continuous improvement* of the OIP, as shown in Table 51. *Depending on the tools for IT-based open innovation*[665] that are used in an OIP, changes are implemented continuously during runtime (no innovation contest is used) or iteratively after the runtime of an OIP (innovation contest is used). Legal restrictions that hinder changing a challenge's rules during runtime have to be considered. Optimization is particularly important for OIPs that run continuously or that are reused, like for instance innovation communities with very long runtimes[666] or standardized OIPs[667]. They are subject to continuous improvements to all areas of the OIP, while OIPs that have a short runtime tend to require bug fixing activities only. To sum it up, OIPs call for frequent iterative improvements, as all cases show. Thus, a corresponding infrastructure to deal with optimizations should be employed.

---

[664] This is in line with Verona et al. (2006), as outlined in Part II.2.3.
[665] See Part II.1.3.
[666] See Part II.1.2.
[667] See Part IV.2.

Table 50: Key quotes concerning the optimize phase

Aspect	Quote
Continuous improvements	*You can see it at this [screenshot of the] prototype. [...] That is how unserAller was developed. A lot of prototypes and trial-and-error. Having a spark and then simply trying it. If it did not work, we leave it at that. That is it. (I_CEO)*
	We try to not break functionality, which is of course not always possible. [...] We try to silently introduce new functionalities. Now and then, there is a new button and you can try it. (I_CIO)
	Many customers have the problem that ideas are submitted twice, five times or ten times. In my eyes, that is not a real problem. Quite the contrary: maybe the ideas will be relevant as a cluster! That's why we, for instance, implemented a calculation of similarity for the customer. (H_CPM)
	We are currently working on a project to automatically group short texts. We have first promising results that we will integrate in the next months to see if that is really useful. (A_CTO)
Depending on the tools for IT-based open innovation	*Due to the law against unfair competition, it is not allowed to extend the length of a competition. [...] That is what we tell our clients as well: once the start and the end of the project's phases are defined, a change is not possible anymore because it is written down in the eligibility requirements. (I_CEO)*
	We have frequent releases once the [internal] contest is live because the customer says we now need this and that feature. Then we implement it accordingly. (H_CPM)
	We have many clients that want to start with a small pilot to try the system. The target is to make it a full version after half a year or one year and really customize it at that point. (H_CPM)
Sources for optimization	*[The client] tells us which functionalities he wants and how he imagines the project. Then we build it the way that it is perfect for him and hope that it is transferrable to our other clients. (I_CEO)*
	We do not always activate the heat map [of Piwik]. We only activate it if we integrate something new in order to evaluate it. (I_CTO)
	What we also do is a biannual user meeting. We invite all customers that are interested. We bring them together to exchange experiences. On the one hand, this exchange is very valuable for the client; on the other hand, we [HIC] get an impression of the bugs and real problems. (H_CPM)
	It is not anonymous. There are many registered users we can ask. In addition, we have a group of beta testers. They are very committed and most of the time, they answer our questions very easily and precisely. Results of them are more certain than when we do any random statistics of traffic data, which is often hard to interpret. (A_CTO)
	We have a list of 15 beta testers. Those are mainly companies, which are completely different. We have software developers who want to test the interfaces, caterers, small producers, large producers and also community members. (I_CEO)

There are multiple *sources for optimization*. Firstly, monitoring of a community often reveals barriers of participation that should be leveled. Furthermore, if innovators of an OIP build up a personal relationship to community managers, it becomes more likely for them to share their wishes. Besides passive reception of ideas, community managers should actively ask for issues and ideas for improvements to the OIP. The cases of Atizo and innosabi show, that an OII can master these approaches by close community management. Besides feedback from the innovators, feedback from organizers should be gathered. One method to do so are offline quality circles, as for instance in the HYVE case, or standardized feedback forms, as for instance in the Atizo case. To sum it up, most valuable feedback is given personally to the OII. Thus, corresponding channels should be considered. Another source for optimization is website tracking. However, tracking results are only a weak hint and should be backed up with qualitative insights.

A sound technological basis facilitates continuous iterative improvements without adding additional complexity to existing code[668], as shown in the innosabi and Atizo case. As the exact impact on the OIP's socio-technical system cannot be determined in advance, rollback options, to revert to prior functionality should be available. These options become relevant for instance if something has not worked out as expected.

The OIIs under investigation employ beta testers in the sense of a perpetual beta paradigm to test optimized functionality. Thereby, they try to make sure, that existing functionality is not changed in its basics which prevents the alteration of existing concepts and, thus, the confusion of innovators. Table 51 summarizes key aspects that have to be considered in the optimize phase of an OIP. The next chapter discusses how OIIs can help organizers to overcome challenges in the lifecycle of OIPs.

Table 51: Major activities in the optimization phase

Activity in the optimization phase
Establish channels for personal feedback by the community and organizers
Rely on personal feedback rather than website tracking
Fix bugs
Continuously improve the OIP's socio-technical system
Facilitate optimization by maintainable and extendable architecture

---

[668] See section 4.1 and section 4.2.

5 Overcoming challenges in the lifecycle of OIPs

This chapter addresses the challenges in an OIP lifecycle as identified in Part II.2.4: selecting the right problems (*section 5.1*), formulating problems (*section 5.2*), overcoming employees' reluctance (*section 5.3*) and facilitating software-mediated knowledge transfer (*section 5.4*).

5.1 Selecting the right problems

The findings of Sieg et al. [669] that companies tend to struggle in selecting appropriate problems that can be solved by innovators, is confirmed. Corresponding quotes are shown in Table 52. While the cases of HYVE and innosabi do not provide any explicit mechanisms to meet this challenge, the case of Atizo provides evidence: OIIs can run workshops to identify an organizer's current problems. In a second step, the problems can be integrated into a solvable problem statement for an OIP. Thus, while using the organizer's expertise on current corporate problems, which might be revealed to the world, the OII can contribute its expertise in selecting those problems that are suitable for an OIP.

Table 52: Key quotes concerning the challenge of selecting the right problems

Challenge	Quote
Selecting the right problems	*There has often been a workshop where we found out in which areas the client has questions. We figured out the questions together with the client and after that, one of us builds the project and checks whether the questions and everything else are correct. (A_CTO)*
	In the first workshop, we actually try to have the most important employees or managers of the client at one table to discuss all of their problems. Not the questions in detail, but just maybe the most important three general problems to find some overall questions concerning these problems. (A_CTO)

However, the selection of an appropriate problem relies on trial-and-error and experience. Terwiesch and Xu [670], for instance, propose a framework to identify suitable problems for a project in the field of open innovation. Applying this or other research findings would enable open innovation intermediaries to select problems based on an informed approach rather than trial-and-error and experience. Having

---

[669] Sieg et al. (2010). See Part II.2.4.
[670] Terwiesch and Xu (2008).

selected a problem, the next challenge is the task formulation, which is tackled in the following section.

5.2   Formulating problems

Formulating appropriate problem statements holds four major challenges: *overcoming a specific language,* helping innovators to *recognize similarities* with their domain, dealing with *linear problem formulation* and formulating problems according to the *ultimate target* [671]. Open innovation intermediaries address these difficulties, as sketched in Table 53.

Table 53: Key quotes concerning the challenge of formulating problems

Challenge	Quote
Overcome specific language	*We recognized that the companies have difficulties formulating good questions. They tend to post too long questions nobody wants to read. The second problem is a language nobody understands. Due to these two problems, we support our clients in formulating a question. […] We are leading edge at this. We do it for four years by now and did by far the most projects. I think we did more than 150 projects with our process. (A_CEO)*
	We have already tested several things and now we have found a mean how the most creative and most understandable questions arise. […] We have a certain framework that we always apply. Firstly, we listen to the clients' problems to explore in which direction the project could go. Secondly, we make three suggestions [for possible questions]. Thirdly, we formulate a question based on the brainstorming we did in the workshop. Finally, this formulation of the question is reviewed and modified by the client. (A_CEO)
Recognize similarities	*We know pretty exactly how we have to ask in order to come to a result. In most cases our client has already a clear idea in which direction it wants to go. So it comes down to finding the formulation of question that fits the imagination and that we can fulfill with our platform so that the community hast fun participating. (I_CEO)*
Linear problem formulation	*Besides marking interesting ideas, which is also a kind of guiding the community, the client can enter an additional text to the question. In this text, he can describe in which direction the ideas should aim. By this, he can influence the task during the runtime of a project because [as said before] he is not allowed to change the question itself. (A_ADM)*
Ultimate target	-

Firstly, OIIs can help organizers to *overcome specific language* and innovators to *recognize similarities* with their domain by taking over the task formulation. By assuming this responsibility, an OII can filter the input of organizers, check if parts of a task are unclear and formulate it according to the innovators' needs. This step can

---

[671] See Part II.2.4.

be incorporated explicitly in a process, as the Atizo and innosabi case show, or be part of a discussion at some point during the design phase, like in the HYVE case. This benefit of an OII that has to understand and formulate the task prior to posting it to an OIP calls for a mediated, rather than a direct task formulation. This again is an argument against the endeavors to provide OIPs as software-as-a-service without the support of the OII, like in the case of Atizo and innosabi. However, a SaaS version might again be viable, if the OII trains an organizer's employees to run the structured workshops as an (organizer-internal) expert, as the case of Atizo shows.

The second issue is the *linear problem formulation*. Besides legal restrictions, revising a problem statement on an OIP is not advisable. However, subtle mechanisms to overcome this problem might be applied, as the case of Atizo shows. Firstly, interesting contributions can be marked in order to direct the innovators towards a desired outcome. By this, a community manager provides hints to the innovators, which resembles an iterative problem formulation and can therefore be seen as a cure for this issue[672]. Secondly, an explanatory text might be added to an innovation project's task formulation during runtime in order to react to frequent misunderstandings on the part of the innovators, as in the Atizo case. Combining these two mechanisms allows to converge towards an iterative problem formulation while complying with legal rules. Finally, the community managers can ensure, that the innovators understands the problem formulation appropriately and actively comment on contributions that do not target an intended outcome, as in the innosabi case. The effort of the three approaches, namely *marking interesting ideas, adding explanatory text* and *active commenting*, increases[673]. Nevertheless, the (assumed) appropriateness to overcome the problem of linear problem formulation increases as well. However, these approaches can only mitigate the issue and not solve it completely.

The issue of formulating problems according to the *ultimate target* remains[674]. Particularly in complex systems, an OII might not be capable to assess all circumstances of a problem and advise the organizer on a more relevant question[675]. In less complex systems, or with a more knowledgeable OII, this issue might be addressed by an iterative problem definition during the process of formulating the task for the OIP.

---

[672] See Part II.2.4.
[673] The least effort approach was mentioned first.
[674] Accordingly, Table 53 does not list any quotes concerning formulating problems according to the ultimate target.
[675] See Part II.2.4.

To sum it up, the issue of problem formulation cannot be solved completely, although all open innovation intermediaries under investigation address this challenge actively. The following section deals with the employees' reluctance towards adapting innovations that stem from an OIP.

5.3   Employees' reluctance

Different approaches can be applied in an OIP lifecycle in order to overcome the employees' reluctance towards the adaption and implementation of ideas that stem from an OIP[676], as shown in Table 54.

Table 54: Key quotes concerning the challenge of employees' reluctance

Aspect	Quote
Activity by the organizer	*An unserAller project involves different departments which ex ante define their solution space and which also take the final decision in each phase. [...] It is a relatively complex process. On the one hand, it has to be transparent and fair for the community, on the other hand, the client needs the opportunity to exert influence on the outcome if you get Pril with chicken flavor. (I_CIO)*
	Due to the fact, that the community generates the ideas and the client provides the knowledge for production, the not-invented-here syndrome can be avoided. It is still the client's product because he himself advanced it that far. He still has to work on the product in order to make it producible. (I_CEO)
	We try to influence the process as much as possible so that there is a wrap-up meeting at the end [of the runtime] where we select the best ideas and create projects from them. Of course, in that point, we reach our limits but we try our very best. (H_PM2)
	It is important to integrate the experts who are busy with the realization later on into the process of the contest. They have to take their time: they have to rate and comment ideas for at least an hour a week with so that they know what they will be facing after the contest. Thus, they also have the feeling of being part of the process. (H_SPM1)
Consequence	*Not only those ideas that have a good evaluation are relevant, but in general those, who have a high activity with a lot of comments, visits or evaluations. (H_CPM)*
	In the beginning, we had projects that did not result in a product [because the process was not yet optimized]. But for instance with Migros, we now have many things that are realized mainly owing to a good process. (A_CEO)

All approaches that are taken to overcome the employees' reluctance require *activity by the organizer*. OIIs can ask an organizer's relevant employees to participate in the innovation project and evaluate interim results, integrate them in workshops and

---

[676] See Part II.2.4.

asks employees to take care of the community management. Table 55 summarizes possible activities of an organizer and outlines which OII applies which approach. The following paragraphs discuss the resulting consequences with regard to effort and the implementation rate of ideas that stem from a respective OIP.

Table 55: Activities to overcome employees' reluctance and its consequences

		innosabi	HYVE	Atizo
Activity by the organzier	Community management		✓	✓
	Frequent interim evaluation	✓		
	Final evaluation		✓	✓
	Improvement workshop			✓
	Implementation workshop			✓
Consequence	Effort	Low	Medium	Medium
	Effect on implementation rate	High	Low	Medium

The data show [677] that most projects that frequently integrate an organizer's employees in a structured process of *frequent interim evaluation* end up in an implemented product, like the innosabi case illustrates[678]. In comparison, approaches that do not apply frequent interim evaluation result in a medium or low implementation rate, like in the Atizo and HYVE case. Consequently, the approach to frequently integrate an organizer's employees in a structured process and give them the opportunity to opt-out ideas they do not like, seems to mitigate the NIH syndrome [679] better than the other approaches. Furthermore, a combination of multiple integration strategies, like for instance in the Atizo case, does not seem to be capable of outrunning the benefits of frequent interim evaluations.

This is particularly interesting, as the required effort that an organizer has to invest for frequent interim evaluations is, compared to the combination of other approaches, particularly low. Thus, it seems promising to continuously integrate an organizer's employees who might opt out contributions in order to overcome the NIH syndrome.

---

[677] See the aspect of consequence in Table 54.
[678] In former times, innosabi committed their clients to implement the innovators' results. This practice is not in place any more. The implementation rate refers to the period since a client is able to decide on the implementation of the innovators' results on its own.
[679] See Part II.2.4.

5.4   Facilitating software-mediated knowledge transfer

In order to ease software-mediated knowledge transfer from the innovators to the organizer, OIIs can create OIPs that offer multiple *channels to submit contributions*[680]. OIPs can allow to upload multiple media (like texts or pictures), and combine online and offline innovation activities by providing innovation toolkits, like the innosabi case shows. Lowering barriers for access has been proven to be a requirement for OIPs, as evident in section 4.1. To achieve this, multiple access modes to an OIP can be offered, like via the OIP's homepage, via Facebook, via mobile devices or via external homepages. Besides that, OIIs should strive for an almost error-free OIP, like in the innosabi and Atizo case.

Table 56: Key quotes concerning the challenge of facilitating software-mediated knowledge transfer

Aspect	Quote
Channels to submit contributions	*He suggested that you can find a code in the application on his homepage and we found this idea amazing. If you know the code, you can get the community yoghurt cheaper. You are allowed to tell the code to your friends like "hey, today the code is 'Frozen-Mary' […]" but after a time friends are stressed out and get their own app to look up the daily code. He also wants to hand out iPads to the queue [in his shop] with that they can have a say, receive the code and get discount on products. These are ideas we never would have figured out by ourselves. (I_CEO)*
	In the snack project, it was partly about the packaging of them: users sent in sketches or only verbal formulations of their ideas. In the latter case we drew these ideas and uploaded the sketches in addition to the text. (I_CEO)
	It is possible to work with a platform on which employees can upload ideas, pictures, texts, tags and other media and where these contributions can be evaluated, commented and criticized. (H_PM1)
Human agents	*We try to influence the process as much as possible so that there is a wrap-up meeting at the end [of the runtime] where we select the best ideas and create projects from them. Of course, in that point, we reach our limits but we try our very best. (H_PM2)*
	[The reports] are directly generated as a spread sheet. Everything is automated with graphics and means so that the client can use it. (A_ADM)
	In the second workshop, we have a clear structure: our workshop cards [with a structured representation of the ideas] can be printed out directly from Atizo. We integrate the best ideas […] In the end, we will have 10 to 15 appropriate and realizable ideas. In the last [third] workshop, whose purpose is the implementation of one of the ideas, we talk about available resources and which part of the managerial board would be the right audience for presenting and realizing the idea. (A_CEO)

---

[680] See Part II.2.4 and Table 56.

Human agents who are to support the software-mediated knowledge transfer, as requested by researchers[681], are represented by community managers, who are established in all cases. Community managers help innovators for instance if problems with the technology arise or if the innovator cannot submit an idea using the offered channels. In addition, OIIs can support an organizer with their expertise and technology to integrate contributions and transfer them to the organizer, as explicitly shown in the Atizo and innosabi case. The present chapter has elaborated how to overcome challenges in the lifecycle of OIPs. The following chapter summarizes the discussion.

---

[681] See Part II.2.4.

6 Summary of Part IV

This part discussed the findings from the three in-depth cases of Part III in a cross-case analysis. Firstly, it was shown that OIIs fulfill the functions of *gatekeeping and brokering, knowledge processing and combination* as well as *demand articulation* for an organizer. Anyhow, OIIs should strive to fulfill complementary functions like *commercialization, scanning and information processing,* as well as *technological services* to extend their portfolio. An additional function *providing and managing technology* was introduced to characterize OIIs further. Secondly, applying the two dimensions of *OIP design* and *OIP management* in that new function of OIIs, a matrix revealed four types of OIP projects: *standardized, facilitated, self-made* and *custom OIP projects*. An OII typically focuses on one type of OIP project. Thirdly, it was shown that the *OIP design process* runs as follows from a socio-technical perspective: Firstly, the *task* has to be fixed in order to determine required *people*, i.e. innovators. Subsequently, the *structure* with a special focus on the motivational system is designed to meet peoples' needs. The structure is finally implemented in the *technology*. This process is a challenge, as an OIP designer can only influence the technical subsystem of an OIP, namely the task and the technology, directly. The next discussion topic outlined the OIP lifecycle model and the key activities in each phase of it. The activities are summarized in Figure 51. Finally, mechanisms that deal with the major challenges in an OIP's lifecycle, i.e. *selecting the right problems, formulating problems, overcoming employees' reluctance* and *facilitating software-mediated knowledge transfer*, were introduced. It was shown, that the challenges can be addressed and leveled to a certain degree.

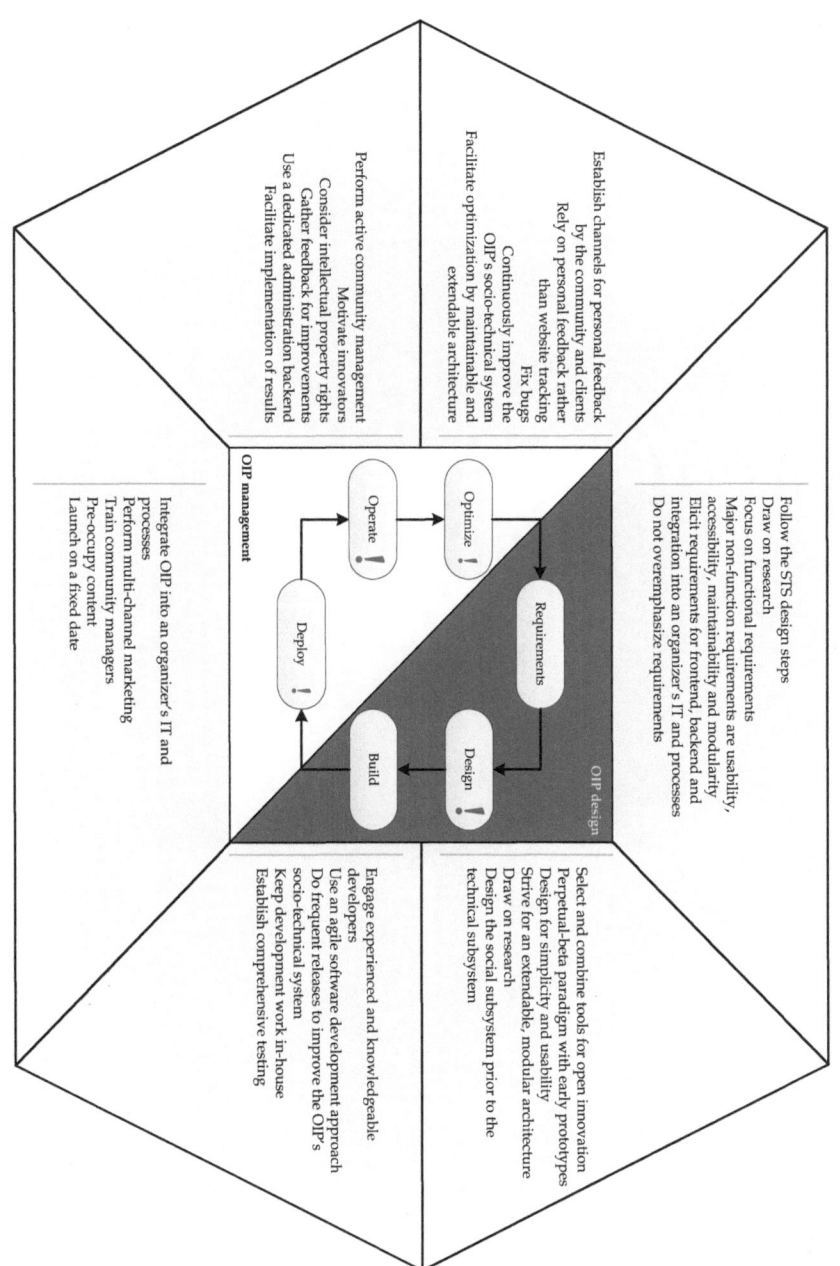

Figure 51: Activities in each phase of the OIP lifecycle model

Part V

Conclusion

1 Summary and contribution

This thesis dealt with managing the lifecycle of OIPs. It provided groundwork in structuring and prioritizing an OIP's lifecycle by developing the OIP lifecycle model. Furthermore, it provided guidelines on what to do in each phase of an OIP's lifecycle and how to deal with challenges that arise. Based on this knowledge, practitioners can now manage the lifecycle of OIPs based on an informed approach. In addition, researchers can answer the initially stated question whether it is more important to take care of a user-centered design process during development, or whether it is more important to invest time in community management[682]. The present chapter summarizes the contributions of *Part I* to *Part V* in order to draw managerial and research implications. It mainly builds on the discussion in Part IV, as outlined in Figure 52.

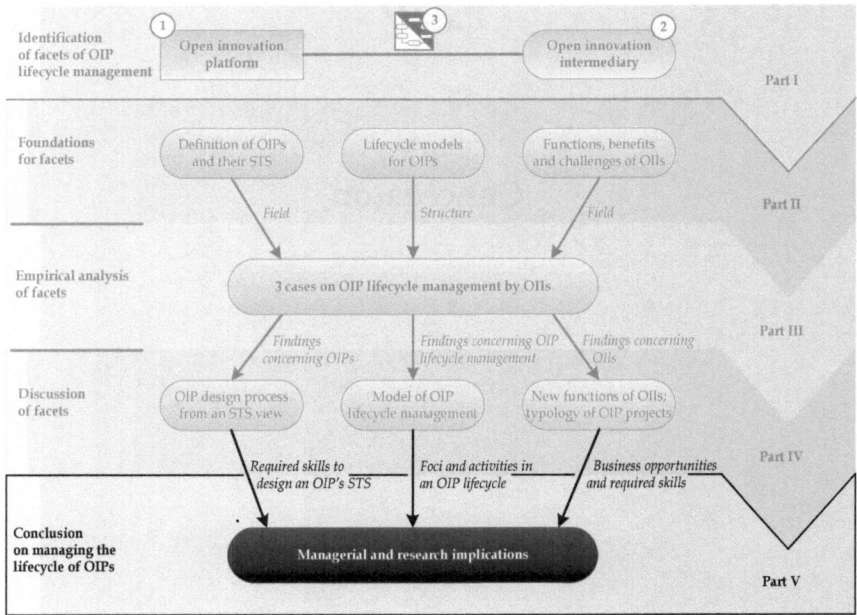

Figure 52: Current progress in the research design

---

[682] See Part I.1. The answer to this sample question can be read from Part IV.4. Both activities are crucial for an OIP's success. However, community management is potentially more time-consuming than designing.

The ensuing *chapters 2* and *3* derive implications from the findings for management and research. While the first provides guidelines for both, organizers and managers of OIPs, the latter chapter aims at stimulating and directing future research in the realm of OIP lifecycle management. Finally, *chapter 4* outlines which parts of this thesis have already been shared with the scientific community.

Before going into the details of each part, the overall structure is reflected. *Part I* introduced the relevance of designing and managing OIPs and the approach to merge currently distinct research streams of open innovation and information systems. The importance of bringing professional software lifecycle management to the management of OIP projects was outlined. *Part II* set the foundations of the relevant facets of designing and managing OIPs. It introduced open innovation intermediaries as professional lifecycle managers and providers of open innovation platforms. The first two topics drew on open innovation literature. Finally, application lifecycle management was described as an approach stemming from information systems literature. It served as a basis for the OIP lifecycle model that structures the activities in an OIP's lifecycle. *Part III* introduced findings from three explorative cases on OIP lifecycle management by established open innovation intermediaries. *Part IV* discussed the findings, along a structure induced by Part II and Part III. Finally, the present *Part V* concludes the thesis, summarizes it on a more abstract level and provides managerial and research implications. The following introduces the major contributions of each part to this thesis.

Part I set the stage by defining the research questions of who should manage an OIP's lifecycle and how this should be done. It thereby outlined the importance of the present research for research and practice and identified the three relevant facets in OIP lifecycle management: (1) the field, i.e. *the OIPs*, (2) the OIP lifecycle managers, i.e. the *open innovation intermediaries*, and (3) an approach to structure the lifecycle of OIPs, i.e. *an OIP lifecycle model*. Against the background of these three facets, the research design was announced using a multiple linear-analytic case study.

Part II introduced *open innovation platforms* as socio-technical systems that incorporate IT-based tools for open innovation. These tools include *innovation contests*, *innovation communities*, *innovation market places* and *innovation toolkits*. Open innovation platforms are used by open innovation intermediaries to integrate innovators into an organizer's innovation process.

Secondly, Part II set foundations for *open innovation intermediaries*, who use OIPs to integrate innovators into an organizer's innovation process. In this setting, the organizer is the OII's client. OIIs can fulfill multiple functions for an organizer, namely the following: *gatekeeping and brokering, middle men between science policy and industry, demand articulation* in the dimension of *connection; knowledge processing and combination, commercialization, foresight and diagnosis* and *scanning and information processing* in the dimension of *collaboration and support;* as well as *intellectual property, testing, validation and training, assessment and evaluation* and *accreditation and standards* in the dimension of *technological services.* Due to the impact of the virtual environment of OIIs, certain challenges arise when working with them. These are *selecting the right problems* that are solvable by externals, *formulating problems* in a way innovators can understand them, *overcoming employees' reluctance,* which might arise due to the NIH syndrome, and *facilitating software-mediated knowledge transfer* in the virtual setting. It was shown that open innovation intermediaries lack an informed approach to design and manage OIPs, which, however, is crucial for an OIP's success.

Due to this nuisance, the third chapter of Part II elaborated on the need for an *OIP lifecycle management* approach and identified application lifecycle management (ALM) as a suitable model to structure the lifecycle of OIPs. ALM provides a *balanced view* and an *appropriate level of abstraction* on the activities of designing and managing an OIP. Furthermore, it is *comprehensive* as it covers all typical activities in a software lifecycle. ALM was adapted to the field of OIPs resulting in the OIP lifecycle model (OIP-LM). According to the OIP-LM, an OIP lifecycle consists of *OIP design* and *OIP management.* Whereas OIP design refers to the phases *requirements, design* and *build,* OIP management refers to the phases *deploy, operate* and *optimize* in an OIP's lifecycle. These foundations served to structure the case analyses which were to come next.

Part III presented findings concerning the approach to design and mange OIPs of three professional OIIs. Therefore, a multiple in-depth holistic *case study,* following a linear-analytic structure, was performed. The three cases were assessed using semi-structured explorative interviews as the main data source. To begin with, results from the OII innosabi GmbH were introduced. innosabi provides an organizer with a standard OIP, i.e. unserAller.de, and takes care of community management. Secondly, the HYVE AG was introduced. This OII builds custom OIPs based on its framework HYVE IdeaNet© and delegates community management to the organizer. Thirdly, the Atizo AG with its OIP Atizo.com was introduced as an OII that runs a standardized, though customizable OIP. At Atizo, community management is shared between the organizer and the OII. For all cases, details on the OII, the OIP and the

OII's lifecycle management were outlined and structured along the foundations of Part II.

Part IV discussed the details given in Part III in a *cross-case analysis*. It showed that OIIs fulfill the *functions* of *gatekeeping and brokering, knowledge processing and combination* as well as *demand articulation*. OIIs thus focus on the dimension of *connecting* organizers to innovators. Anyhow, they should strive to fulfill complementary functions like *commercialization, scanning and information processing, intellectual property, testing, validation and training, assessment and evaluation* and *accreditation and standards. Providing and managing technology* was introduced as an additional function since it constitutes a core task for OIIs.

OIP design and *OIP management* were identified as dimensions in the function *providing and managing technology. OIP management* can either be the OII's or the organizer's responsibility, *OIP design* is either standardized or custom. The two dimensions defined four types of OIP projects: *standardized, facilitated, self-made* and *custom OIP projects*. Benefits and drawbacks of each project type were discussed. An OII typically focuses on one type of OIP project, which allows selecting OIIs according to the type of OIP projects they run.

The findings summarized so far allow answering research question I) concerning *who* should manage the lifecycle of OIPs. The following paragraphs summarize findings that help to answer research question II) regarding *how* OIP lifecycle management should be done.

Designing an OIP's socio-technical system should follow a certain *OIP design process*: An OII has to fix the *task*, i.e. the problem formulation, first in order to determine the required *people*, i.e. the innovators. Subsequently, the *structure*, such as the motivational system, is designed to meet the needs of the innovators. Finally, the structures are implemented in the *technology*, e.g. the functions and graphical design of an OIP. This process is a challenge, as an OIP designer can only directly influence the technical subsystem (i.e. task and technology), but not the social subsystem (i.e. people and structure), which, nevertheless, is crucial for an OIP's success.

Finally, the model to manage an OIP's lifecycle, namely the *OIP lifecycle model* was elaborated. In this model, crucial activities in an OIP lifecycle were identified. OIIs should build on experience and research collaboration in the *requirements phase*. However, no particular skills are required here, as default techniques to elicit and manage requirements in agile software development approaches are applied. The subsequent *design phase* is particularly important in the lifecycle of an OIP, as designing a meaningful socio-technical system is challenging. The OIP design process mentioned in the paragraph above should be applied. A sound architecture of the OIP facilitates scalability for an OII. Compared to traditional web-design projects, the *build phase* loses importance. It was outlined that the technical subsystem has to function well from an innovator's perspective in order to represent the intended social subsystem and not destroy the innovators' motivation. According to the observations of this thesis, the activities in the *deployment phase* are shifting from technical deployment towards marketing activities, which aim at recruiting potential innovators. This phase is important, but not as crucial as the design and operate phase. In the *operation phase*, innovators have to be motivated to participate in the OIP and support is necessary. In standard OIPs, *optimization* is also important, though not a crucial phase in the overall process.

In addition to the OIP lifecycle model, mechanisms that deal with the following challenges in an OIP's lifecycle were introduced: *selecting the right problems, formulating problems, overcoming employees' reluctance* and *facilitating software-mediated knowledge transfer*. It was shown that the challenges can be addressed and leveled to a certain degree.

In addition, it was outlined that the skills required to design and manage OIPs differ from those that are necessary to design a default website. While default web-design agencies focus on the design and build phase, OIIs should focus on the design and operate phase and have expertise in deployment and optimization.

Part V concludes this thesis with a short summary of its contributions and derives managerial and research implications. The following chapter proceeds with deriving managerial implications.

2 Managerial implications

This thesis focused on how to manage the lifecycle of open innovation platforms. In order to perform a sound OIP lifecycle management, knowledge in the topics of designing and managing open innovation platforms is required. Three actors are mainly relevant to OIP lifecycle management and therefore considered in the managerial implications: *Organizers,* as initiators of OIPs, *OIP lifecycle managers,* as designers and managers of OIPs, and *open innovation intermediaries* as professional OIP lifecycle managers. In the following, implications are derived for all three, *organizers,* who need to select an appropriate *OIP lifecycle manager* as well as for *OIIs,* who need to perform a sound management of an OIP's lifecycle.

Implications for organizers

When an *organizer* intends to build an OIP, they should look for a specialized open innovation intermediary rather than a default web-design agency[683]. A well designed and managed OIP reduces the risk of a social media disaster like in the Henkel case[684]. Furthermore, OIIs should help an organizer to deal with the challenges of *selecting the right problems, formulating problems, overcoming employees' reluctance* and *facilitating software-mediated knowledge transfer.*

In order to select an appropriate OII, an organizer has to decide whether it wants to run a standard or custom OIP and whether OIP management should be performed by the OII or by themselves. The two dimensions define four OIP project types as shown in Figure 53. OIIs typically focus on one of these OIP project types. If an organizer plans an OIP project, they should select an OII that specializes in that particular OIP project type. The following paragraphs outline the implications of the four OIP project types.

---

[683] The case of designing and managing an OIP in-house is not discussed in this thesis. In that case, an organizer's employees should meet the same requirements and follow the same recommendations as OIIs.

[684] For details on the Henkel case see Part I.1 and Spiegel Online (2011).

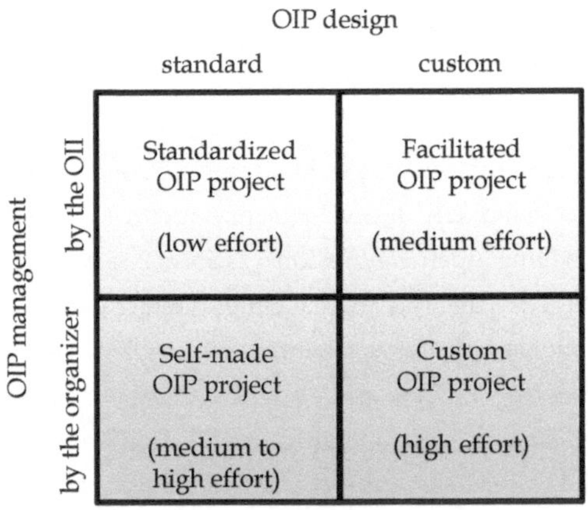

OIP design

	standard	custom
by the OII	Standardized OIP project (low effort)	Facilitated OIP project (medium effort)
by the organizer	Self-made OIP project (medium to high effort)	Custom OIP project (high effort)

OIP management

Figure 53: Classification of OIP projects and effort for an organizer

While a *custom OIP* design yields the benefits of a branded graphical design and corresponding marketing opportunities, it requires effort and knowledge in designing an OIP. Setting up a *standard OIP* is quicker, easier and allows drawing on a tested design and, in most cases, an existing community of innovators, which eases attracting relevant innovators. In OIP management, community management accounts for most effort and requires profound skills. On the one hand, getting this done professionally *by the OII*, yields the benefit of low effort for the organizer and an informed community management. On the other hand, taking care of OIP management by oneself (*by the organizer*) allows for deeper insights into the innovators[685].

Bringing the two dimensions together, a *standardized OIP project* yields the benefit of a low effort project for an organizer in a standardized process and OIP. innosabi runs typical projects of this type[686]. *Facilitated OIP projects* draw on an OII's full service by running a project to the specifications of the organizer. No particular knowledge in the field is required. An organizer can run a *self-made OIP project* without the help of an OII, for instance by using a SaaS OIP and performing the community management in-house. This actually requires skills to manage the OIP. Finally, *custom OIP projects* require the most profound knowledge, as this is the most individual project type allowing for the strongest influence and customization but yields the benefits of a close integration of the innovators. HYVE is an OII that

---

[685] For more benefits and drawbacks see Table 30 on page 153.
[686] For details see Part III.2.

specializes on this type of OIP project[687]. Besides these types of OIP projects, in-between types exist that combine benefits of the types. Atizo runs such projects with a shared community management and standard, though customizable OIP[688].

Based on the typology of OIP projects, an organizer can select an appropriate OII that can run a project according to their objectives. If the organizer for instance wants to run an OIP project without much effort or lacks the necessary knowledge, they should choose an OII that specialized on standardized OIP projects. innosabi for instance runs primarily standardized OIP projects and is thus a potentially suitable OII for the organizer seeking this type of OIP project. If an organizer is looking for a rather deep integration with the innovators and a branded OIP, they should address an OII that masters custom OIP projects like HYVE. The present investigation classified three OIIs according to the type of OIP project they primarily run, as shown in Figure 55. An organizer has to select what kind of OIP project it intends to run and can choose, based on this, a suitable OII which is specialized on that type of OIP project. Additional OIIs that an organizer can draw on are listed in Part III.1.2.

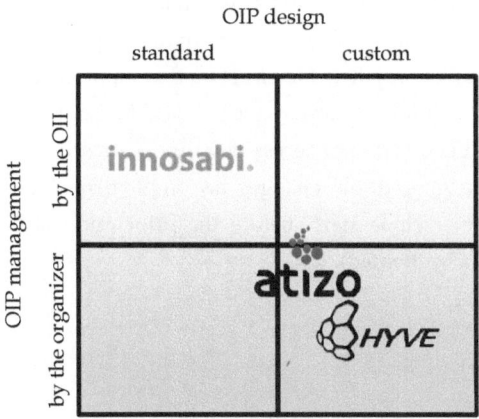

Figure 54: Classification of OIIs according to type of OIP project

As the present paragraphs have shed light on the implications for organizers, the following derives implications for OIP lifecycle managers in general and for OIIs in particular.

---

[687] For details see Part III.3.
[688] For details see Part III.4.

Implications for OIP lifecycle managers

Open innovation intermediaries are major players in designing and managing an OIP's lifecycle. This thesis provides insights into the core activities in an OIP's lifecycle, which allows OIIs to rely on an *informed approach* on OIP design and management rather than on trial-and-error. The following summarizes implications for OIP lifecycle managers in general[689] and thereafter in particular for OIIs that take care of professional OIP lifecycle management organizers.

The OIP lifecycle model[690] structures the activities in an OIP's lifecycle along the six phases *requirements, design, build, deploy, operate* and *optimize*. While the phases *design* and *operate* are particularly important in an OIP's lifecycle and should thus be focused on, the phases *requirements* and *build* are losing importance in comparison to traditional web-design projects. The phases *deploy* and *optimize* are also important in an OIP lifecycle, yet not as crucial as the phases design and operate. Activities in each phase differ from traditional web-design projects. Corresponding success factors in each phase of the OIP lifecycle model are shown in Figure 55.

Although the OIP lifecycle model resembles a plan-driven software lifecycle model, OIP lifecycle managers should apply an agile approach to *run* OIP projects. The OIP lifecycle model only helps to *structure* the required activities. The design of an OIP, i.e. the OIP's socio-technical system, should be improved incrementally. The phases in an OIP's lifecycle do not necessarily run consecutively, but can overlap due to parallel circles (e.g. multiple changes are implemented at the same time) or iterations (e.g. another circle starts before the prior one finished or two or more process steps have to be iterated).

---

[689] As OIIs are OIP lifecycle managers, these implications are valid for them as well.
[690] See Figure 55 and Part IV.5.

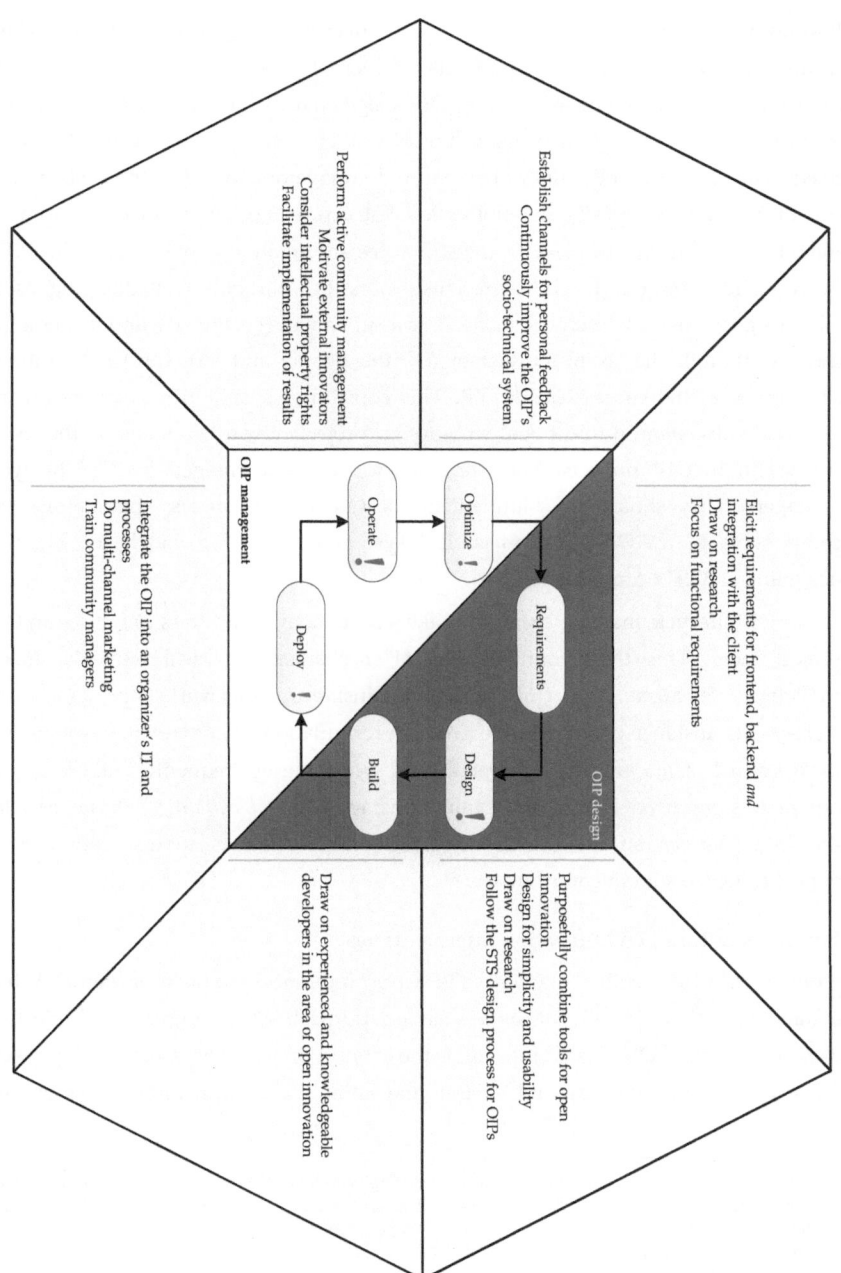

Figure 55: Success factors in each phase of the OIP lifecycle model

Besides managing the overall process, an OIP lifecycle manager has to develop skills in designing and managing in particular the social subsystem of OIPs in order to attract and motivate innovators[691]. The *OIP design process*[692] that is to achieve this goal follows *four major steps*: *Firstly*, the task of an OIP and thereby the objective has to be fixed. This in turn determines the required innovators to solve the task, which constitutes *step two*. *Thirdly*, the motivational structures need to be designed to meet the innovators' needs. In *step four*, the structures are finally implemented in the OIP's functionalities and graphical design. This process is a particularly challenging, as an OIP designer can only influence the technical subsystem (i.e. the task and technology) directly, though the social subsystem (i.e. the people and structure) substantially contributes to the success of an OIP. This constitutes a shift from a focus on the technical subsystem in traditional web-design projects towards a focus on the social subsystem in OIP projects. This shift has to be acknowledged by OIP lifecycle managers. They should thus implement the OIP design process that defines the components of an OIP's socio-technical system in order to match the logic of designing an OIP's components.

OIP lifecycle managers should make sure to actively address *challenges* an OIP project faces. This thesis outlines several mechanisms to deal with the major challenges, i.e. *selecting the right problems*, for instance using workshops; *formulating problems*, for instance by an iterative problem formulation; overcoming an organizer's *employees' reluctance* and the connected NIH syndrome by frequently integrating an organizer's employees in interim evaluations; as well as *facilitating software-mediated knowledge transfer* on an OIP by offering multiple access modes and various opportunities to submit contributions[693].

Implications for open innovation intermediaries

From an OII's perspective, the *type of OIP project* influences the business model. While a standard *OIP design* allows for a scalable approach when running multiple OIP projects, a custom OIP design intensifies the connection to the organizer. OIP projects are getting more complex and custom the latter way. Accordingly, custom OIP

---

[691] This is in line with the claim of Verona et al. (2006), who state that *creating incentive systems* is an important skill of OIIs. Accordingly, *profiling customers* is also a core skill of OIIs, so that they can define which innovators are relevant to solve a task. However, *tracking customers* is not a required core skill of OIIs (opposing to Verona et al. 2006). See also Part II.2.3.

[692] See Part IV.3.

[693] For details see Part IV.1.

projects require comprehensive skills in designing and managing OIPs from the OIIs. If *OIP management* is taken care of by the organizer, this again allows for scalability at the OII, though the quality of OIP management is not granted. If an OII takes care of community management on its own, a standard quality can be granted, although effort might be high. Nonetheless, a proper community management is crucial for an OIP's success (for instance the output quality and quantity) and thus for an organizer's satisfaction. Thus, when an organizer wants to perform community management on its own, a dedicated training to educate the community managers should be conducted.

According to the OIP project types[694], it seems that OIIs have to choose, whether they want to exploit the benefits of scalability (like in the innosabi case) or individuality (like in the HYVE case) in OIP projects. The Atizo case, however, shows how to masters balancing the two extremes by providing a hybrid approach. An OII can reach scalability in customizable OIP projects by providing a sound technological basis that is easily adaptable to an organizer's specification. Thus, OIIs can jointly optimize the seemingly opposing dimensions of scalability and individuality of OIP projects.

Summarizing this, an *OII's success* is influenced by three major aspects, which can be presented in a formula[695]: *Knowledge on open innovation* as well as *basic technical skills* are mandatory to run OIP projects. If at least one of them is low or not present (reaches zero), the overall success of an OIP's project is jeopardized (the overall term aspires towards zero). That is the reason for their multiplicative relationship. In order to allow for scalability and consequently facilitate growth, an OII should strive for a *superior technical design* of the OIP. This is an additive relationship, as this factor is optional for an OII's success. If it is amiss, an OII can still be successful, but not grow to its full potential.

Figure 56: Success formula for open innovation intermediaries

---

[694] See Part IV.2.
[695] See Figure 56.

OIIs currently fulfill multiple *functions* for organizers[696]. Especially in OIP design, OIIs rise towards their potential and cover relevant functions. Yet, compared to functions of traditional innovation intermediaries[697], OIIs fall back in services, which could be offered during OIP management. This leaves room for OIIs to differentiate themselves from their competitors and build up a competitive advantage. OIIs could offer services to organizers that fulfill the functions *commercialization* of an OIP's results, managing *intellectual property* rights for an organizer, *testing, validation and training, assessing and evaluating technology, accreditation and standards* setting or *regulation and arbitration,* as highlighted in Table 57.

Table 57: Functions of open innovation intermediaries during OIP design and management

Dimension	Function	OIP design	OIP management
Connection	Gatekeeping and brokering	✓	✓
	Demand articulation	✓	✓
Collaboration and support	Knowledge processing and combination	✓	✓
	Commercialization	-	o
Technological services	Intellectual property	✓	o
	Testing, validation and training	-	o
	Assessment and evaluation	-	o
	Accreditation and standards	-	o
	Regulation and arbitration	-	o
	Providing and managing technology	✓	✓

o complementary function OIIs could fulfill
- function not recommended for OIIs

In sum, this thesis shows that designing and managing an OIP is far more challenging than designing and managing a default web-design project. Therefore, an organizer of an OIP should acquire a professional OII that holds the required skills and offers the required functions – as outlined above – to manage an OIP lifecycle. Having outlined managerial implications in this chapter, the pursuing chapter outlines paths for further research endeavors.

---

[696] See Table 57.
[697] See Part II.2.2.

3   Research implications

This thesis set the foundations to structure and prioritizes the design and management of open innovation platforms by merging research on open innovation and
information systems. Thereby, contributions were made to both fields. IS research
benefits from additional areas of application for application lifecycle management,
which built the basis for the OIP lifecycle model, and cases thereon. Open innovation
research benefits from a professionalization of OIP lifecycle management and a prioritization of design and management activities in an OIP lifecycle. The conducted case
study revealed basic mechanisms that influence the lifecycle of an OIP as well as tasks
that arise from this. The most pressing challenges in an OIP lifecycle, as identified in
prior research, were addressed and mechanisms to solve them were outlined.
Anyhow, by its explorative character the case study unraveled needs for further
research endeavors. This chapter is structured along the topics discussed in Part IV.
Table 58 summarizes the resulting research questions, which are explained below.

Table 58: Directions for future research

Topic	Research question
Functions of OIIs	Which functions should OIIs add to their service?
	- Impact of type of OIP on meaningful configuration of functions
	- Value proposition and competitive advantage
Typology of OIP projects	What is the impact of the OIP project type on OIP success?
	- Elicit cases for all OIP project types
	- Extend typology
OIP design process	How can an active design of an OIP's social subsystem be realized?
	- Links of the social subsystem and the technical subsystem
	- Methods to design the social subsystem of OIPs
Lifecycle of OIPs	What are the major influences on appropriate OIP lifecycle activities?
	- Required and competitive advantage-building activities
	- Tools supporting these activities
Challenges in an OIP's lifecycle	Which methods are most effective to overcome challenges in an OIP lifecycle?
	- Identification and adaption of existing methods
	- Development of new methods

Functions of open innovation intermediaries

Part IV.1 identified functions that might complement an OII's services. Although there is a great deal of overlapping functions the OIIs under investigation fulfill, the data also show that some OIIs fulfill functions others do not fulfill. Future research should thus investigate the reasons for this finding. I suppose, that the type of OIP, as identified for instance in Hallerstede et al.[698], influences the offered functionalities. This would explain why Innocentive, as an OII that focuses on professional innovators[699], provides services concerning intellectual property rights[700], while other OIIs, like the three under investigation, do not offer this in their context of predominantly hobbyists and tinkerers [701] . Taking one further step, certain combinations of functions might add more value for organizers than others. This example shows that there are several types of OIIs, which are not covered by the data of this thesis. Thus, future research should investigate more OIIs in order to derive meaningful configurations of OII's services and factors influencing them. The results of this line of research might be valuable new business models for OIIs.

As discussed in Part IV.1 as well, the function of *knowledge processing and combination* is fulfilled by the community and not the OII. The OII rather provides an infrastructure to facilitate the process of knowledge processing and combination. The infrastructure can, therefore, be reproduced rather easily. The core competency of internal (intransparent) knowledge processing and combination, that traditional innovation intermediaries used to exploit[702], was comparably hard to copy. This constitutes a major change in the business model of traditional innovation intermediaries compared to open innovation intermediaries. The impact on a sustainable competitive advantage is still unclear and remains to be explored.

Typology of OIP projects

Part IV.2 explores four types of OIP projects. The present thesis covers only cases for two of the four types, namely custom and standardized OIP projects, as the typology was not known before. In order to complete findings on the typology, cases of OIIs that perform facilitated and self-made OIP projects should be gathered for further

---

[698] Hallerstede et al. (2010).
[699] See Part II.1.3.3.
[700] See www.innocentive.com/for-solvers/intellectual-property; retrieved September 23, 2012 and Lichtenthaler and Ernst (2008).
[701] Hallerstede et al. (2010).
[702] Howells (2006).

verification and detailed explanation of the proposed typology. With this knowledge, superior OIP project types for a certain aim might become identifiable: I assume, for instance, that running custom OIP projects is not a viable business model for open innovation intermediaries that strive for scalability, as the basic functions of an OIP remain and should thus be standardized (though being configurable). In addition, the impact of the OIP type on the output of an OIP, i.e. for instance the quality of an OIP's results, is unknown. Is it necessary to run a custom OIP in order to generate ground-breaking ideas? On the other hand, in the dimension of OIP management, the more and more popular offering of software-as-a-service OIPs enable organizers to perform their own OIP management. The impact of professional versus amateur OIP management has not been quantified yet, though prior research already showed that OIP management is crucial for an OIP's success[703]. The concurrent processing of innovation projects managed by the OII innosabi, and innovation projects in the new SaaS-version of unserAller, which are managed by the organizer, offers a unique opportunity to investigate the influence of professional experience on OIP management.

Another convincing direction for research towards a typology of OIIs and their OIPs was already proposed by Verona et al.[704]. They call for a typology along two dimensions: the type of knowledge an OII creates and the phase of the innovation process that an OII focuses on. As outlined in the case of innosabi, an OII can support all three phases of the innovation process, including implementation of results. The other two OIIs do not support implementation. This is a particular interesting topic, as I assume that now that the concept is proven to work, more approaches to integrate innovators in the implementation phase will arise in the context of OIPs. The typology of Verona et al.[705] could help to understand the extent of the support of the implementation phase and thus might help to structure a possible future market of OIIs.

OIP design process

Part IV.3 discusses a design process for OIPs and identifies important relations among the components of an OIP's STS. However, details of the relations between the social subsystem and the technical subsystem of OIPs are still unknown. In addition,

---

[703] See for instance Adamczyk (2012).
[704] Verona et al. (2006).
[705] Verona et al. (2006).

details of the influence of each component on the overall success of an OIP have not been clarified so far. It is for instance unclear how and to what extent the structure influences the technology. Making the relations explicit would help OIP designers to understand the relationships and allow for an informed design of an OIP[706]. Based on this knowledge, future research could derive explicit methods to design the components of an OIP's STS. A special focus should be set on the social subsystem, as experiences and insights are still scarce in this area, while experience in designing the technical subsystem already exists to a certain extent. This endeavor would complement the current line of research on designing OIPs by for instance Boudreau et al.[707]. They look into details of the social subsystem, more precisely the number of participants who are needed in order to optimize an OIP's performance.

Lifecycle of OIPs

Multiple phases in the lifecycle of an OIP were considered important[708]. However, the data do not reveal mastering which of the phases constitutes a sustainable competitive advantage for OIIs, nor does it reveal threshold points that have to be overcome in order to fulfill a minimum requirement in a particular phase to grant an OIP's success. With this knowledge, OIIs would be able to focus on using their limited resources towards a best performing OIP whilst simultaneously building a competitive advantage.

Furthermore, the data revealed the importance of different foci in the lifecycle of an OIP, depending on the type of OIP. In addition, activities in some phases of the OIP lifecycle model differ based on the type of OIP. As of now, internal versus external OIPs are supposed to have a significant impact on the activities in an OIP's lifecycle. Additional aspects influencing an OIP's lifecycle management are still unknown and should therefore be explored. This would allow deriving a contingency approach for activities in an OIP's lifecycle. Thus, the OIP lifecycle model would be refined. This endeavor would complement the line of research to derive a typology of OIIs as proposed above. For both, the typology of OIIs and the contingency approach of OIPs, identifying important influencing factors on OIP lifecycle management is required. This allows making use of synergies.

---

[706] See Bostrom and Heinen (1977).
[707] Boudreau, Lacetera and Lakhani (2011).
[708] See Part IV.4.

A so far neglected topic in the lifecycle of OIPs is administration backends for OIP. There is some research on how to manage OIPs, for instance the carrying out of community management, but there is no insight into tools that support these processes and their impact on success [709]. Nevertheless, the data show that administration backends are crucial for the execution of a proper community management. Consequently, this line of research is promising to improve the overall quality of an OIP's output.

Challenges in an OIP's lifecycle

This study revealed challenges in an OIP lifecycle and mechanisms how the OIIs under investigation deal with these challenges[710]. This represents a first step towards answering the question how to address the outlined challenges. However, the cases show only a limited set of mechanisms to deal with the challenges. Especially task formulation is a neglected field since the 1970s articles of Hackman and Lawler[711] and considerations on task complexity in the 1980s by various authors[712]. However, current research shows that the specificity of the task formulation has a strong impact on the required OIP design[713]. Thus, future research should identify or, if necessary, develop additional mechanisms to address the challenges and evaluate how suitable each mechanism is with regard to this. Future endeavors could for example draw on other research disciplines like marketing to address these questions. Kaulio[714] for instance shows that different problems require different mechanisms to integrate innovators in new product development. His logic might be adapted to the field of OIPs. In addition, marketing research offers insights into formulating custom marketing messages for certain audiences[715], which might be conducive to formulating tasks.

In sum, the fields of open innovation platforms and open innovation intermediaries remain to be rather unexplored in the presented level of abstraction. This thesis provided groundwork and a structuring of the lifecycle of OIPs and thereby revealed plenty of opportunities for exiting future research. The following chapter outlines how parts of this research have been communicated to the research community up to now.

---

[709] See Adamczyk (2012).
[710] See Part II.2.4 and Part IV.1.
[711] Hackman and Lawler (1971).
[712] For instance Campbell (1988).
[713] Hallerstede and Bullinger (2010); Piller and Walcher (2006).
[714] Kaulio (1998).
[715] Gooley and Lattin (2000).

4   Communication of research

Research aims at the publication of findings in order to discuss them with the research community and make them available to practitioners[716]. A common mean to do so is publishing in conferences and journals. As claimed in Part III.1.5, foundations and results of the single cases were discussed with the OIIs and the research community in order to validate and improve them. Table 59 lists previous publications that revealed parts of this thesis and the major overlaps with it.

Table 59: Previous publications with content from this thesis

Publication	Relevant content	Integrated in
Hallerstede, Neyer, Bullinger and Moeslein (2010)	Foundations: Innovation contests	Part II.1
	Foundations: Design elements of OIPs	Part II.1
Hallerstede and Bullinger (2010)	Foundations: Design elements of OIPs	Part II.1
Hallerstede, Danzinger, Bullinger and Moeslein (2011)	Foundations: Adaption of ALM as a framework for the lifecycle of OIPs	Part II.3
Hallerstede, Bullinger and Moeslein (2012a)	Foundations: Innovation intermediaries	Part II.1
	Empirical data: Processes on innosabi's OIP	Part III.2
Hallerstede, Bullinger and Moeslein (2012b)	Foundations: Innovation communities	Part II.1
	Empirical data: OIP lifecycle management at innosabi	Part III.2
Hallerstede and Bullinger (2012)	Foundations: Socio-technical systems theory	Part II.1
	Foundations: Application lifecycle management	Part II.3
	Empirical data: OIP lifecycle management at HYVE	Part III.3
Hallerstede (2012)	Foundations: Functions of innovation intermediaries	Part II.1
	Discussion: Functions of OIIs	Part IV.1

The major contribution to each paper remains with the first author, who is also the author of this thesis. The present thesis incorporates feedback of researchers and practitioners concerning each publication. Thus, all chapters that integrate content from prior publications represent improved versions.

---

[716] Rocco and Hatcher (2011).

References

Abrahamsson, P., Salo, O., Ronkainen, J., & Warsta, J. (2002). *Agile software development methods*. Espoo: Vtt Publications.

Adamczyk, S. (2012). *Managing Innovation Contests: Challenges of Attraction and Facilitation*. Dissertation, University of Erlangen-Nuremberg.

Adamczyk, S., Boehler, D., Bullinger, A. C., & Moeslein, K. M. (2011). Facilitating interaction in web-based communities: the case of a community for innovation in healthcare. In H.-U. Heiss, P. Pepper, H. Schlingloff, & J. Schneider (Eds.), *Informatik 2011. Informatik schafft Communities. Lecture Notes in Informatics - Proceedings, Series of the Gesellschaft fuer Informatik* (p. 228).

Adamczyk, S., Bullinger, A. C., & Moeslein, K. M. (2010). Call for attention – attracting and activating innovators. *Proceedings of the R&D Management Conference 2010* (pp. 1–14). Manchester.

Adamczyk, S., Bullinger, A. C., & Moeslein, K. M. (2012). Innovation Contests: A Review, Classification and Outlook. *Creativity and Innovation Management, 21*(4), 335–360.

Addicks, J. S., & Steffens, U. (2008). Supporting landscape dependent evaluation of enterprise applications. *Multikonferenz Wirtschaftsinformatik 2008* (pp. 1815–1825). München.

Afuah, A. (2003). Redefining Firm Boundaries in the Face of the Internet: Are Firms Really Shrinking? *Academy of Management Review, 28*(1), 34–53.

Afuah, A., & Tucci, C. L. (2012). Crowdsourcing As a Solution to Distant Search. *The Academy of Management Review, 37*(3), 355–375.

Ahmad, R., Li, Z., & Azam, F. (2005). Web engineering: a new emerging discipline. *Proceedings of the IEEE International Conference on Emerging Technologies* (pp. 445–450). Islamabat.

Ahuja, G. (2000). Collaboration Networks, Structural Holes, and Innovation: A Longitudinal Study. *Administrative Science Quarterly, 45*(3), 425–455.

Andrews, A. A., & Pradhan, A. S. (2001). Ethical Issues in Empirical Software Engineering: The Limits of Policy. *Empirical Software Engineering, 6*, 105–110.

Andrews, D., Preece, J., & Turoff, M. (2002). A conceptual framework for demographic groups resistant to online community interaction. *Proceedings of the 34th Annual Hawaii International Conference on System Sciences* (pp. 9–24). Maui, HI.

Antikainen, M. J., & Vaataja, H. K. (2010). Rewarding in open innovation communities – how to motivate members. *International Journal of Entrepreneurship and Innovation Management, 11*(4), 440–456.

Arlow, J., & Neustadt, I. (2005). *UML 2 and the Unified Process: Practical Object-Oriented Analysis and Design* (2nd ed.). Boston, MA: Addison-Wesley.

Armstrong, A., & Hagel III, J. (1996). The Real Value of on-line Communities. *Harvard Business Review,* (3), 134–141.

Arora, A., Fosfuri, A., & Gambardella, A. (2004). *Markets for Technology: The Economics of Innovation and Corporate Strategy.* Cambridge, MA: The MIT Press.

Bakos, J. Y. (1997). Reducing Buyer Search Costs: Implications for Electronic Marketplaces. *Management Science, 43*(12), 1676–1692.

Bartl, M., & Ivanovic, I. (2010). Netnography – Finding the right balance between automated and manual research. In P. Brauckmann (Ed.), *Web-Monitoring: Gewinnung und Analyse von Daten über das Kommunikationsverhalten im Internet* (1st ed., pp. 157–174). Konstanz: UVK Verlagsgesellschaft mbH.

Bass, L., Clements, P., & Kazman, R. (2003). *Software Architecture in Practice* (2nd ed.). Boston, MA: Addison-Wesley.

Bateman, P. J., Gray, P. H., & Butler, B. S. (2010). The Impact of Community Commitment on Participation in Online Communities. *Information Systems Research, 22*(4), 841–854.

Beck, K. (1999). Embracing change with extreme programming. *IEEE Computer, 32*(10), 70–77.

Beck, K., & Andres, C. (2004). *Extreme Programming Explained: Embrace Change* (2nd ed.). Harlow: Addison-Wesley.

Becker, B., & Gassmann, O. (2006). Gaining leverage effects from knowledge modes within corporate incubators. *R&D Management, 36*(1), 1–16.

Beedle, M., Devos, M., Sharon, Y., Schwaber, K., & Sutherland, J. (1999). SCRUM: An extension pattern language for hyperproductive software development. *Pattern Languages of Program Design, 4,* 637–651.

Beenen, G., Ling, K., Wang, X., Chang, K., Frankowski, D., Resnick, P., & Kraut, R. E. (2004). Using social psychology to motivate contributions to online communities. In J. Herbsleb & G. Olson (Eds.), *Proceedings of the 2004 ACM conference on Computer supported cooperative work* (Vol. 10, pp. 212–221). New York, NY.

Behkamal, B., Kahani, M., & Akbari, M. K. (2009). Customizing ISO 9126 quality model for evaluation of B2B applications. *Information and Software Technology, 51*(3), 599–609.

Bell, G. (2009). *Building Social Web Applications.* Sebastopol, CA: O'Reilly.

Bellagio, D. E., & Milligan, T. J. (2005). *Software Configuration Management Strategies and IBM Rational ClearCase* (2nd ed.). Armonk, NY: IBM Press.

Belz, F.-M., & Baumbach, W. (2010). Netnography as a Method of Lead User Identification. *Creativity and Innovation Management, 19*(3), 304–313.

Benassi, M., & Di Minin, A. (2009). Playing in between: patent brokers in markets for technology. *R&D Management, 39*(1), 68–86.

Benbasat, I., Goldstein, D. K., & Mead, M. (1987). The Case Research Strategy in Studies of Information Systems. *MIS Quarterly, 11*(3), 369–386.

Benediktsson, O., Dalcher, D., & Thorbergsson, H. (2006). Comparison of software development life cycles: a multiproject experiment. *IEE Proceedings - Software, 153*(3), 87–101.

Bessant, J., & Rush, H. (1995). Building bridges for innovation: the role of consultants in technology transfer. *Research Policy, 24*(1), 97–114.

Bhaskar, R. (2008). *A Realist Theory of Science.* New York, NY: Routledge.

Biemans, W. G. (1991). User and third-party involvement in developing medical equipment innovations. *Technovation, 11*(3), 163–182.

Bishop, M. (2009). *The Total Economic Impact TM Of InnoCentive Challenges - Single Company Case Study.* Cambridge, MA: Forrester Research.

Bjoerkdahl, J., & Linder, M. (2010). Formulating problems for commercializing new technologies: The case of greening. *Paper presented at the Druid Summer Conference 2010* (pp. 1–40). London.

BMWi. (2011). BMWi-Pressemitteilungen: Rösler prämiert erfolgreichste IKT-Unternehmensgründungen. Retrieved January 14, 2012, from http://www.bmwi.de/BMWi/Navigation/Presse/pressemitteilungen,did=4062 52.html.

Boehm, B. (1979). Guidelines for verifying and validating software requirements and design specifications. *Proceedings of the European Conference of Applied Information Technology* (pp. 711–719). London.

Boehm, B. (1988). A spiral model of software development and enhancement. *IEEE Computer, 21*(5), 61–72.

Boehm, B., Egyed, A., Kwan, J., Port, D., Shah, A., & Madachy, R. (1998). Using the WinWin spiral model: a case study. *IEEE Computer, 31*(7), 33–44.

Boehm, B. W. (1981). *Software engineering economics*. Englewood Cliffs, NJ: Prentice-Hall.

Boehm, B. W., & Turner, R. (2003). *Balancing agility and discipline: a guide for the perplexed*. Boston, MA: Addison-Wesley.

Bogers, M., Afuah, A., & Bastian, B. (2010). Users as Innovators: A Review, Critique, and Future Research Directions. *Journal of Management, 36*(4), 857–875.

Bon, J. Van, Pondman, D., & Kemmerling, G. (2002). *It Service Management*. Zaltbommel: Van Haren Publishing.

Boon, W., & Moors, E. (2008). Demand articulation in intermediary organisations: The case of orphan drugs in the Netherlands. *Technological Forecasting and Social Change, 75*(5), 644–671.

Borchardt, A., & Goethlich, S. E. (2007). Erkenntnisgewinnung durch Fallstudien. In S. Albers, D. Klapper, U. Konradt, A. Walter, & J. Wolf (Eds.), *Methodik der empirischen Forschung* (Vol. 2, pp. 33–48). Wiesbaden: Gabler.

Borst, I. (2010). *Understanding Crowdsourcing-Effects of motivation and rewards on participation and performance in voluntary online activities*. Rotterdam: ERIM.

Bosch, J. (2000). *Design and Use of Software Architectures*. Harlow: Addison-Wesley.

Bostrom, R. P., & Heinen, J. S. (1977). MIS problems and failures: a socio-technical perspective - Part I: The causes. *MIS Quarterly, 1*(3), 17–32.

Boudreau, K. J., Lacetera, N., & Lakhani, K. R. (2011). Incentives and Problem Uncertainty in Innovation Contests: An Empirical Analysis. *Management Science, 57*(5), 843–863.

Boudreau, K., Lacetera, N., & Lakhani, K. R. (2008). Incentives versus diversity: Re-examining the link between competition and innovation. *Proceedings of the Wharton Technology Conference* (pp. 1–31). Philadelphia, PA.

Brabham, D. C. (2009). Moving the Crowd at Threadless: Motivations for Participation in a Crowdsourcing Application. *Paper presented at the AEJMC conference* (pp. 1–16). Chicago, IL.

Brabham, D. C. (2010). Moving the Crowd At Threadless. *Information Communication Society, 13*(8), 1122–1145.

Braun, D. (1993). Who Governs Intermediary Agencies? Principal-Agent Relations in Research Policy-Making. *Journal of Public Policy, 13*(02), 135–162.

Braun, D., & Guston, D. H. (2003). Principal-agent theory and research policy: an introduction. *Science and Public Policy, 30*(5), 302–308.

Brown, J. S., & Duiguid, P. (1991). Organizational Learning and Communities of Practice: Toward a Unified View of Working, Learning and Innovation. *Organizational Science, 2*(1), 40–57.

Broy, M., & Rausch, A. (2005). Das neue V-Modell® XT. *Informatik-Spektrum, 28*(3), 220–229.

Bryl, V., Giorgini, P., & Mylopoulos, J. (2009). Designing socio-technical systems: from stakeholder goals to social networks. *Requirements Engineering, 14*(1), 47–70.

Bullinger, A. C., Haller, J. B. A., & Moeslein, K. M. (2009). Innovation Mobs – Unlocking the Innovation Potential of Virtual Communities. *Proceedings of the Fifteenth Americas Conference on Information Systems* (pp. 1–8). San Francisco, CA.

Bullinger, A. C., Hoffmann, H., & Leimeister, J. M. (2011). The next step – open prototyping. *Proceedings of the European Conference of Information Systems* (pp. 1–12). Helsinki.

Bullinger, A. C., & Moeslein, K. M. (2011). Innovation Contests: Systematization of the Field and Future Research. *International Journal of Virtual Communities and Social Networking, 3*(1), 1–12.

Bullinger, A. C., Neyer, A.-K., Rass, M., & Moeslein, K. M. (2010). Community-based innovation contests: Where cooperation meets competition. *Creativity & Innovation Management, 19*(3), 290–303.

Burke, R., Rangaswamy, A., & Gupta, S. (2001). Rethinking market research in the digital world. In J. Wind & V. Mahajan (Eds.), *Digital Marketing: Global strategies from the world's leading experts* (pp. 226–255). New York, NY: John Wiley & Sons.

Burt, R. (1995). *Structural Holes: The Social Structure of Competition.* Cambridge MA: Harvard University Press.

Butler, B., Sproull, L., Kiesler, S., & Kraut, R. (2007). Community effort in online groups: Who does the work and why? *Carnegie Mellon University Research Showcase Paper 90* (pp. 1–32). Pittsburgh, PA: Human-Computer Interaction Institute.

Callon, M. (1980). The state and technical innovation: a case study of the electrical vehicle in France. *Research Policy, 9*(4), 358–376.

Callon, M. (1994). Is Science a Public Good? *Science, Technology & Human Values, 19*(4), 395–424.

Campbell, D. (1988). Task complexity: A review and analysis. *Academy of management review, 13*(1), 40–52.

Cao, L., & Ramesh, B. (2008). Agile requirements engineering practices and challenges: an empirical study. *IEEE Software, 25*(1), 60–67.

Carlsson, B., & Jacobsson, S. (1997). Diversity creation and technological systems: a technology policy perspective. In C. Edquist (Ed.), *Systems of innovation technologies institutions and organizations* (pp. 266–294). Oxon: Pinter.

Carmel, E. (1999). *Global Software Teams: Colloborating Across Borders and Time Zones.* Saddle River, NJ: Prentice Hall.

Carmel, E., Whitaker, R. D., & George, J. F. (1993). PD and joint application design: a transatlantic comparison. *Communications of the ACM, 36*(6), 40–48.

Carvalho, A. (2009). In search of excellence - Innovation contests to foster innovation and entrepreneurship in Portugal. *CEFAGE-UE Working Paper* (pp. 1–16). Évora: Universidade de Évora.

Cash, D. (2001). In order to aid diffusing useful and practical information: boundary organizations and agricultural extension. *Science, Technology & Human Values, 26*(4), 431–453.

Cherns, a. (1976). The Principles of Sociotechnical Design. *Human Relations, 29*(8), 783–792.

Chesbrough, H. (2006a). Open Innovation: A New Paradigm for Understanding Industrial Innovation. In H. Chesbrough, W. Vanhaverbeke, & J. West (Eds.), *Open Innovation: Researching a New Paradigm.* Oxford: Oxford University Press.

Chesbrough, H. W. (2003). *Open Innovation: The New Imperative for Creating and Profiting from Technology.* Boston, MA: Harvard Business Press.

Chesbrough, H. W. (2006b). *Open business models: how to thrive in the new innovation landscape.* Boston, MA: Mcgraw-Hill Professional.

Christensen, C. M. (2006). *The innovator's dilemma.* New York, NY: Harper Business.

Christensen, C. M., & Bower, J. L. (1996). Customer Power, Strategic Investment, and the Failure of Leading Firms. *Strategic Management Journal, 17*(3), 197–218.

Clegg, C. W. (2000). Sociotechnical principles for system design. *Applied ergonomics*, *31*(5), 463–77.

Cockburn, A. (2008). Using both incremental and iterative development. *STSC CrossTalk*, *21*(5), 27–30.

Collier, A. (1994). *Critical Realism: An Introduction to Roy Bhaskar's Philosophy*. London: Verso.

Conradi, R., & Westfechtel, B. (1998). Version models for software configuration management. *ACM Computing Surveys*, *30*(2), 232–282.

Constantine, L. L., & Lockwood, L. A. D. (2002). Usage-centered engineering for Web applications. *IEEE Software*, *19*(2), 42–50.

Corbin, J., & Strauss, A. (2008). *Basics of Qualitative Research: Techniques and Procedures for Developing Grounded Theory* (3rd ed.). Thousand Oaks, CA: Sage Publications Inc.

Cothrel, J., & Williams, R. L. (1999). On-line communities: helping them form and grow. *Journal of Knowledge Management*, *3*(1), 54–60.

Craincross, F. (1997). *The death of distance: How the communication revolution will change our lives* (Vol. 6). Boston, MA: Harvard Business School Press.

Crainer, S. (2001). *The Tom Peters Phenomenon: Corporate Man to Corporate Skunk*. Oxford: Capstone.

Creswell, J. W. (2007). *Qualitative inquiry and research design*. Thousand Oaks, CA: Sage Publications Inc.

Crumlish, C., & Malone, E. (2009). *Designing Social Interfaces - Princilpes, Patterns, and Practices for Improving the User Experience*. Sebastopol, CA: O'Reilly.

Cummings, T. G., & Srivastva, S. (1977). *Management of Work - A socio-technical Systems Approach*. Kent, OH: Kent State University Press.

Czarnitzki, D., & Spielkamp, A. (2000). Business services in Germany: bridges for innovation. *ZEW Discussion Paper 00-52*. Mannheim: ZEW.

Daft, R. L., Lengel, R. H., & Trevino, L. K. (1987). Message Equivocality, Media Selection, and Manager Performance: Implications for Information Systems. *MIS Quarterly*, *11*(3), 355.

Dahan, E., & Hauser, J. R. (2002). The virtual customer. *Journal of Product Innovation Management*, *19*(5), 332–353.

Dahlander, L., & Gann, D. M. (2010). How open is innovation? *Research Policy, 39*(6), 699–709.

Dahlander, L., & Wallin, M. W. (2006). A man on the inside: Unlocking communities as complementary assets. *Research Policy, 35*(8), 1243–1259.

Dalcher, D. (2002). Life cycle design and management. In M. Stevens (Ed.), *Project Management Pathways* (pp. 60.2–30). High Wycombe: APM Press.

Dalziel, M. (2010). Why do innovation intermediaries exist? *Paper presented at the Druid Summer Conference 2010* (pp. 1–23). London.

Datta, P. (2007). An Agent-Mediated Knowledge-in-Motion Model. *Journal of the Association for Information Systems, 8*(5), 287–311.

Davidsen, M. K., & Krogstie, J. (2010). A longitudinal study of development and maintenance. *Information and Software Technology, 52*(7), 707–719.

Davis, A. M. (1992). Operational prototyping: A new development approach. *IEEE Software, 9*(5), 70–78.

Davis, F. D., & Venkatesh, V. (2004). Toward Preprototype User Acceptance Testing of New Information Systems: Implications for Software Project Management. *IEEE Transactions on Engineering Management, 51*(1), 31–46.

Dearle, A. (2007). Software Deployment, Past, Present and Future. *Future of Software Engineering* (pp. 269–284). Minneapolis, MN.

Denzin, N. K. (2009). *The Research Act: A Theoretical Introduction to Sociological Methods.* Piscataway, NJ: Aldine Transaction.

Diener, K., & Piller, F. (2010). *The Market for Open Innovation - Increasing the efficiency and effectiveness of the innovation process. A Market Study of Open Innovation Intermediaries.* Raleigh, NC: Lulu.

Dix, A., Finlay, J. E., Abowd, G. D., & Beale, R. (2004). *Human-Computer Interaction.* Upper Saddle River, NJ: Pearson Education.

Downes, L., & Mui, C. (1998). *Unleashing the Killer App: Digital Strategies for Market Dominance.* Boston, MA: Harvard Business School Press.

Droeschel, W., & Wiemers, M. (1999). *Das V-Modell 97.* München: Oldenbourg Wissenschaftsverlag.

Dubé, L., & Paré, G. (2003). Rigor in Information Systems Positivist Case Research: Current Practices, Trends, and Recommendations. *MIS Quarterly, 27*(4), 597–636.

Ebert, C. (2008). *Systematisches Requirements Engineering und Management*. Heidelberg: Dpunkt.Verlag GmbH.

Ebner, W., Leimeister, J. M., Bretschneider, U., & Krcmar, H. (2008). Leveraging the Wisdom of Crowds: Designing an IT-supported Ideas Competition for an ERP Software Company. *Proceedings of the 41st Annual Hawaii International Conference on System Sciences* (p. 417). Hawaii, HI.

Ebner, W., Leimeister, J. M., & Krcmar, H. (2010). Community Engineering for Innovations: The Ideas Competition as a method to nurture a Virtual Community for Innovations. *R&D Management, 40*(4), 342–356.

Elbaum, S., Rothermel, G., Karre, S., & Fisher II, M. (2005). Leveraging user-session data to support Web application testing. *IEEE Transactions on Software Engineering, 31*(3), 187–202.

Elliott, G. (2004). *Global business information technology: an integrated systems approach*. Harlow: Addison Wesley.

Emery, F. E., & Trist, E. L. (1960). Socio-technical Systems. In C. W. Churchman & M. Verhulst (Eds.), *Management Sciences - Models and Techniques. Proceedings of the sixth international Meeting of the Institute of Management Science* (pp. 83–97). Paris.

Erlikh, L. (2000). Leveraging legacy system dollars for e-business. *IT Professional, 2*(3), 17–23.

Ernst, H. (2004). Virtual customer integration–Maximizing the impact of customer integration on new product performance. In S. Albers & K. Brockhoff (Eds.), *Cross-functional innovation management* (pp. 191–208). Wiesbaden: Gabler.

Escalona, M. J., & Koch, N. (2003). Requirements engineering for web applications: a comparative study. *Journal of Web Engineering, 2*(3), 193–212.

Evans, P., & Wurster, T. S. (1999). *Blown to Bits: How the New Economics of Information Transforms Strategy*. Boston, MA: Harvard Business School Press.

Fahey, L., & Prusak, L. (1998). The eleven deadliest sins of knowledge management. *California Management Review, 40*(3), 265–276.

Farnie, D. A. (1979). *The English Cotton Industry and World Market, 1815-96*. Oxon: Oxford University Press.

Flanagan, J. C. (1954). The critical incident technique. *Psychological bulletin, 51*(4), 327–358.

Flyvbjerg, B. (2006). Five Misunderstandings About Case-Study Research. *Qualitative Inquiry, 12*(2), 219–245.

Fowler, M. (2002). *Patterns of Enterprise Application Architecture.* Amsterdam: Addison-Wesley.

Fowler, S., & Stanwick, V. (2004). *Web Application Design Handbook: Best Practices for Web-Based Software.* San Francisco, CA: Morgan Kaufmann.

Fox, W. (1995). Sociotechnical system principles and guidelines: past and present. *The Journal of applied behavioral science, 31*(1), 91–105.

Franke, N., & Von Hippel, E. (2003). Satisfying heterogeneous user needs via innovation toolkits: the case of Apache security software. *Research Policy, 32*(7), 1199–1215.

Fraternali, P. (1999). Tools and approaches for developing data-intensive Web applications: a survey. *ACM Computing Surveys, 31*(3), 227–263.

Fueller, J. (2006). Why Consumers Engage in Virtual New Product Developments Initiated by Producers. *Advances in Consumer Research, 33*(1), 639–646.

Fueller, J., Bartl, M., Ernst, H., & Muehlbacher, H. (2006). Community based innovation: How to integrate members of virtual communities into new product development. *Electronic Commerce Research, 6*(1), 57–73.

Fueller, J., Matzler, K., & Hoppe, M. (2008). Brand community members as a source of innovation. *Journal of Product Innovation Management, 25*(6), 608–619.

Fueller, J., Schmid, M., Hutter, K., Hautz, J., Gebauer, J., & Kuhn, M. (2009). What motivates and hinders employees to engage in internal innovation communities? *Proceedings of the 2nd ISPIM Innovation Symposium* (pp. 1–17). New York, NY.

Galitz, W. O. (2007). *The essential guide to user interface design: an introduction to GUI design principles and techniques* (3rd ed.). Indianapolis: Wiley.

Garlan, D., & Shaw, M. (1993). An Introduction to Software Architecture. In V. Ambriola & G. Tortora (Eds.), *Advances in Software Engineering and Knowledge Engineering* (pp. 1–39). River Edge, NJ: World Scientific Publishing Company.

Gassmann, O. (2006). Opening up the innovation process: towards an agenda. *R&D Management, 36*(3), 223–228.

Gassmann, O. (2010). *Crowdsourcing - Innovationsmanagement mit Schwarmintelligenz.* München: Hanser Fachbuchverlag.

Gassmann, O., & Enkel, E. (2004). Towards a theory of open innovation: three core process archetypes. *Paper presented at the R&D management conference* (pp. 1–18). Sesimbra.

Gassmann, O., Enkel, E., & Chesbrough, H. (2010). The future of open innovation. *R&D Management, 40*(3), 213–221.

Gassmann, O., Sandmeier, P., & Wecht, C. H. (2006). Extreme customer innovation in the front-end: learning from a new software paradigm. *International Journal of Technology Management, 33*(1), 46–66.

Geels, F. W. (2004). From sectoral systems of innovation to socio-technical systems. *Research Policy, 33*(6-7), 897–920.

Gerner, J., Naramore, E., Owens, M. L., & Warden, M. (2006). *Professional LAMP: Linux, Apache, MySQL, and PHP Web Development.* Indianapolis, IN: Wrox Press.

Gladwell, M. (2000). *The Tipping Point: How Little Things Can Make a Big Difference.* Boston, MA: Little, Brown and Company.

Glass, R. L. (2002). Searching for the holy grail of software engineering. *Communications of the ACM, 45*(5), 15–16.

Gooley, C. G., & Lattin, J. M. (2000). Dynamic Customization of Marketing Messages in Interactive Media. *Stanford Research Paper Series Research Paper No. 1664* (pp. 1–33). Stanford, CA: Graduate School of Business Stanford University.

Greer, C. R., & Lei, D. (2012). Collaborative Innovation with Customers: A Review of the Literature and Suggestions for Future Research. *International Journal of Management Reviews, 14*(1), 63–84.

Gulati, R., & Gargiulo, M. (1999). Where Do Interorganizational Networks Come From? *American Journal of Sociology, 104*(5), 1439–1438.

Guntamukkala, V., Wen, H., & Tarn, J. (2006). An empirical study of selecting software development life cycle models. *Human Systems Management, 25*(4), 265–278.

Guston, D. H. (1996). Principal-agent theory and the structure of science policy. *Science and Public Policy, 23*(4), 229–240.

Guston, D. H. (1999). Stabilizing the Boundary between US Politics and Science: The Role of the Office of Technology Transfer as a Boundary Organization. *Social Studies of Science, 29*(1), 87–111.

Hackman, J., & Lawler, E. (1971). Employee reactions to job characteristics. *Journal of applied psychology, 55*(3), 259–286.

Haegerstrand, T. (1952). *The propagation of innovation waves.* Lund: Royal University of Lund, Dept. of Geography.

Hagel III, J., & Rayport, J. F. (1997). The new infomediaries. *McKinsey Quarterly, 4*, 54–70.

Hall, H., & Graham, D. (2004). Creation and recreation: motivating collaboration to generate knowledge capital in online communities. *International Journal of Information Management, 24*(3), 235–246.

Haller, J. B. A. (2012). *Open Evaluation*. Dissertation, University of Erlangen-Nuremberg.

Haller, J. B. A., Bullinger, A. C., & Moeslein, K. M. (2011). Innovationswettbewerbe. *Wirtschaftsinformatik, 53*(2), 105–108.

Haller, J. B. A., Neyer, A.-K., & Bullinger, A. C. (2009). Beyond the Black Box of Idea Contests. *Paper presented at the EURAM Annual Conference 2009* (pp. 1–37). Liverpool.

Hallerstede, S. H. (2012). A classification of Open Innovation intermediaries. *Paper presented at the Research Seminar Innovation & Value Creation* (pp. 1–6). Chemnitz.

Hallerstede, S. H., & Bullinger, A. C. (2010). Do you know where you go? A taxonomy of online innovation contests. *Proceedings of the XXIth ISPIM Conference* (pp. 1–12). Bilbao.

Hallerstede, S. H., & Bullinger, A. C. (2012). Managing the Lifecycle of Online Innovation Contests – A Case Study. *Proceedings of the XXIIIth ISPIM Conference* (pp. 1–13). Barcelona.

Hallerstede, S. H., Bullinger, A. C., & Moeslein, K. M. (2012a). Community-basierte Open Innovation von der Suche bis zur Implementierung – der Fall des Innovationsintermediärs innosabi. *Proceedings der Multikonferenz Wirtschaftsinformatik 2012* (pp. 1735–1746). Braunschweig.

Hallerstede, S. H., Bullinger, A. C., & Moeslein, K. M. (2012b). Design and Management of Web-Based Innovation Communities: A Lifecycle Approach. *Proceedings of the Eighteenth Americas Conference on Information Systems* (pp. 1–10). Seattle, WA.

Hallerstede, S. H., Danzinger, F., Bullinger, A. C., & Moeslein, K. M. (2011). Akzeptanzorientiertes Application Lifecycle Management. *HMD - Praxis der Wirtschaftsinformatik, 278*, 30–40.

Hallerstede, S. H., Neyer, A.-K., Bullinger, A. C., & Moeslein, K. M. (2010). Normalo? Tüftler? Profi? Eine Typologisierung von Innovationswettbewerben. *Proceedings der Multikonferenz Wirtschaftsinformatik 2010* (pp. 1229–1240). Göttingen.

Hansen, M. T., Chesbrough, H. W., Nohria, N., & Sull, D. N. (2000). Networked incubators. Hothouses of the new economy. *Harvard Business Review*, *78*(5), 74–84, 199.

Hargadon, A. B. (1998). Firms as knowledge brokers. *California Management Review*, *10*(3), 209–227.

Hargadon, A. B., & Sutton, R. I. (1997). Technology brokering and innovation in a product development firm. *Administrative Science Quarterly*, *42*(4), 718–749.

Harhoff, D. (2003). Profiting from voluntary information spillovers: how users benefit by freely revealing their innovations. *Research Policy*, *32*(10), 1753–1769.

Hass, A. M. J., & Hass, G. (2003). *Configuration Management Principles and Practice*. Amsterdam: Addison-Wesley.

Henkel, J. (2006). Selective revealing in open innovation processes: The case of embedded Linux. *Research Policy*, *35*(7), 953–969.

Hickey, A. M., & Davis, A. M. (2002). Requirements elicitation and elicitation technique selection: a model for two knowledge-intensive software development processes. *Proceedings of the 36th Annual Hawaii International Conference on System Sciences* (pp. 96–105). Hawaii, HI.

Hill, C. (1967). *Reformation to Industrial Revolution*. London: Weidenfeld & Nicholson.

Von Hippel, E. (2005). *Democratizing Innovation*. Cambridge MA: MIT Press.

Von Hippel, E., Franke, N., & Pruegl, R. (2009). Pyramiding: Efficient search for rare subjects. *Research Policy*, *38*(9), 1397–1406.

Von Hippel, E., & Katz, R. (2002). Shifting Innovation to Users via Toolkits. *Management Science*, *48*(7), 821–833.

Hirsig, C., & Hirschmann, T. (2010). Atizo: Untersützung durch Produkt-, Dienstleistungs- und Marketingideen. In O. Gassmann (Ed.), *Crowdsourcing - Innovationsmanagement mit Schwarmintelligenz* (pp. 73–90). München: Hanser Fachbuchverlag.

Hochstein, A., Zarnekow, R., & Brenner, W. (2004). ITIL als Common-Practice-Referenzmodell für das IT-Service-Management - Formale Beurteilung und Implikationen für die Praxis. *Wirtschaftsinformatik*, *46*(5), 382–389.

Hoehn, R., & Hoeppner, S. (2008). *Das V-Modell XT: Grundlagen, Methodik und Anwendungen*. Berlin: Springer.

Hoffmann, M., & Beaumont, T. (1997). *Application Development: Managing the Project Life Cycle*. Carlsbad, CA: Mc Press.

Holck, J. (2003). 4 Perspectives on Web Information Systems. *Proceedings of the 36th Annual Hawaii International Conference on System Sciences* (p. 265.2). Washington, DC.

Hossain, M. (2012). Performance and Potential of Open Innovation Intermediaries. *Procedia - Social and Behavioral Sciences, 58*, 754–764.

Howells, J. (1999). Research and Technology Outsourcing and Innovation Systems: an Exploratory Analysis. *Industry & Innovation, 6*(1), 111 – 129.

Howells, J. (2006). Intermediation and the role of intermediaries in innovation. *Research Policy, 35*(5), 715–728.

Hoyer, V., Schroth, C., Stanoevska-Slabeva, K., & Janner, T. (2007). Web 2.0-Entwicklung – ewige Beta-Version. *HMD–Praxis der Wirtschaftsinformatik, 255*, 78–87.

Hrastinski, S., Kviselius, N. Z., Ozan, H., & Edenius, M. (2010). A Review of Technologies for Open Innovation: Characteristics and Future Trends. *Proceedings of the 43rd Hawaii International Conference on System Sciences* (pp. 1–10). Hawaii, HI.

Hueck, S., Fueller, J., Bartl, M., & Leckenwalter, R. (2008). Community Research: Analyse von Online-Communities im Rahmen der Produktentwicklung bei Gore. In H. Kaul & C. Steinmann (Eds.), *Community Marketing* (pp. 187–199). Stuttgart: Schäffer-Poeschel.

Huizingh, E. K. R. E. (2011). Open innovation: State of the art and future perspectives. *Technovation, 31*(1), 2–9.

Huston, L., & Sakkab, N. (2006). Connect and Develop: Inside Procter & Gamble's New Model for Innovation. *Harvard Business Review, 58*, 1–11.

Hutter, K., Hautz, J., Fueller, J., Mueller, J., & Matzler, K. (2011). Communitition: The Tension between Competition and Collaboration in Community-Based Design Contests. *Creativity and Innovation Management, 20*(1), 3–21.

IEEE. (1990). *IEEE Standard Glossary of Software Engineering Terminology*. New York, NY: IEEE.

Ihlenburg, D. (2011). *Interaktionsplattformen und Kundenintegration in Industriegütermärkten*. Wiesbaden: Gabler.

Iriberri, A., & Leroy, G. (2009). A life-cycle perspective on online community success. *ACM Computing Surveys, 41*(2), 11.1–11.29.

Jaervi, K., Schallmo, D. R. A., & Kutvonen, A. (2011). The Business of Open Innovation Intermediaries. *Proceedings of the XXIIth ISPIM Conference* (pp. 1–15). Hamburg.

Jahn, T. (2005). Marktplatz für ideen. *McK Wissen, 15*, 82–87.

Jain, R. (2010). Investigation of Governance Mechanisms for Crowdsourcing Initiatives. *Proceedings of the Sixteenth Americas Conference of Computer Information Systems* (pp. 1–8). Lima.

Jawecki, G., & Fueller, J. (2008). How to use the innovative potential of online communities? Netnography – an unobtrusive research method to absorb the knowledge and creativity of online communities. *International Journal of Business Process Integration and Management, 3*(4), 248–255.

Jeppesen, L. B. (2005). User Toolkits for Innovation: Consumers Support Each Other. *Journal of Product Innovation Management, 22*(4), 347–362.

Jeppesen, L. B., & Lakhani, K. R. (2010). Marginality and Problem-Solving Effectiveness in Broadcast Search. *Organization Science, 21*(5), 1016–1033.

Jones, C. (1996). *Applied Software Measurement: Assuring Productivity and Quality.* New York, NY: Mcgraw-Hill.

Katz, R., & Allen, T. J. (1982). Investigating the Not Invented Here (NIH) syndrome: A look at the performance, tenure, and communication patterns of 50 R&D Project Groups. *R&D Management, 12*(1), 7–20.

Kaulio, M. A. (1998). Customer, consumer and user involvement in product development: A framework and a review of selected methods. *Total Quality Management, 9*(1), 141–149.

Kelly, S. E. (2003). Public Bioethics and Publics: Consensus, Boundaries, and Participation in Biomedical Science Policy. *Science, Technology, & Human Values, 28*(3), 339–364.

Kernighan, B. W., & Plauger, P. J. (1976). *Software Tools.* Harlow: Addison-Wesley Professional.

Keyes, J. (2004). *Software Configuration Management.* Boca Raton, FL: Auerbach Publications.

Kim, A. J. (2000). *Community Building on the Web.* Berkeley, CA: Peachpit Press.

Klein, D., & Lechner, U. (2009). The Ideas Competition as Tool of Change Management – Participatory Behaviour and Cultural Perception. *Proceedings of the XXth ISPIM Conference* (pp. 1–14). Wien.

Klerkx, L., & Leeuwis, C. (2008). Matching demand and supply in the agricultural knowledge infrastructure: Experiences with innovation intermediaries. *Food Policy, 33*(3), 260–276.

Klerkx, L., & Leeuwis, C. (2009). Establishment and embedding of innovation brokers at different innovation system levels: Insights from the Dutch agricultural sector. *Technological Forecasting and Social Change, 76*(6), 849–860.

Kodama, T. (2008). The role of intermediation and absorptive capacity in facilitating university–industry linkages—An empirical study of TAMA in Japan. *Research Policy, 37*(8), 1224–1240.

Koh, J., Kim, Y.-G., Butler, B., & Bock, G.-W. (2007). Encouraging participation in virtual communities. *Communications of the ACM, 50*(2), 69–74.

Komus, A., & Wauch, F. (2008). *Wikimanagement: Was Unternehmen von Social Software und Web 2.0 lernen können.* Oldenbourg Wissenschaftsverlag.

Kordon, F., & Luqi. (2002). An Introduction to Rapid System Prototyping. *IEEE Transactions on Software Engineering, 28*(9), 817–821.

Kotonya, G., & Sommerville, I. (1998). *Requirements Engineering: Processes and Techniques.* New York, NY: Wiley.

Kozinets, R. V. (1999). E-Tribalized Marketing? The Strategic Implications of Virtual Communities of Consumption. *European Management Journal, 17*(3), 252–264.

KPMG (2001). *Knowledge Management im Kontext von eBusiness - Status quo und Perspektiven 2001.* Berlin: KPMG Knowledge Management.

Krasner, G. E., & Pope, S. T. (1988). A cookbook for using the model-view controller user interface paradigm in Smalltalk-80. *Journal of Object-Oriented Programming, 1*(3), 26–49.

Krcmar, H. (2010). *Informationsmanagement* (5th ed.). Heidelberg: Springer.

Krishna, S., Sahay, S., & Walsham, G. (2004). Managing Cross-cultural issues in global software outsourcing. *Communications of the ACM, 47*(4), 62–66.

Kruchten, P. (2003). *The Rational Unified Process: An Introduction* (3rd ed.). Boston, MA: Addison-Wesley.

Larman, C., & Basili, V. R. (2003). Iterative and Incremental Development: A brief history. *IEEE Computer, 36*(6), 47–56.

Lave, J., & Wenger, E. (1991). *Situated Learning: Legitimate Peripheral Participation.* Cambridge MA: Cambridge University Press.

Lee, D.-H., In, H. P., Lee, K., Park, S., & Hinchey, M. (2009). A Survival Kit: Adaptive Hardware/Software Codesign Life-Cycle Model. *IEEE Computer, 42*(2), 100–102.

Leimeister, J., & Krcmar, H. (2004). Revisiting the virtual community business model. *Proceedings of the Tenth Americas Conference on Information Systems* (pp. 2716–2726). New York, NY.

Leimeister, J. M., Huber, M., Bretschneider, U., & Krcmar, H. (2009). Leveraging Crowdsourcing: Activation-Supporting Components for IT-Based Ideas Competition. *Journal of Management Information Systems, 26*(1), 197–224.

Van Lente, H., Hekkert, M., Smits, R., & Van Waveren, B. (2003). Roles of systemic intermediaries in transition processes. *International Journal of Innovation Management, 7*(3), 247–279.

Lichtenthaler, U., & Ernst, H. (2008). Innovation Intermediaries: Why Internet Marketplaces for Technology Have Not Yet Met the Expectations. *Creativity and Innovation Management, 17*(1), 14–25.

Lichtenthaler, U., & Ernst, H. (2009). Opening up the innovation process: the role of technology aggressiveness. *R&D Management, 39*(1), 38–54.

Linder, J. C., Jarvenpaa, S., & Davenport, T. H. (2003). Toward an innovation sourcing strategy. *MIT Sloan Management Review, 44*(4), 43–50.

Lindsay, V. J. (2004). Computer-assisted qualititative data analysis: application in an export study. In R. Marschan-Piekkari & C. Welch (Eds.), *Handbook of qualitative research methods for international business* (pp. 486–506). Cheltham: Edward Elgar.

Loeliger, J., & McCullough, M. (2012). *Version Control with Git: Powerful tools and techniques for collaborative software development* (2nd ed.). Sebastopol, CA: O'Reilly.

Lopez-Vega, H., & Vanhaverbeke, W. (2009). Connecting open and closed innovation markets: A typology of intermediaries. *MPRA Paper No. 27017.* München: Munich Personal RePEc Archive.

Lowe, D., & Henderson-Sellers, B. (2001). Characteristics of web development processes. *Proceedings of the SSGRR-2001: Infrastructure for E-Business, E-Education, and E-Science* (pp. 1–12). L'Aquila.

Luethje, C., & Herstatt, C. (2004). The Lead User method: an outline of empirical findings and issues for future research. *R&D Management, 34*(5), 553-568.

Lynn, L., Aram, J. D., & Reddy, N. M. (1997). Technology communities and innovation communities. *Journal of Engineering & Technology Management, 14*, 129-145.

Lynn, L. H., Mohan Reddy, N., & Aram, J. D. (1996). Linking technology and institutions: the innovation community framework. *Research Policy, 25*(1), 91-106.

Macaulay, L. A. (1996). *Requirements Engineering*. London: Springer.

Maes, P. (1999). Smart Commerce: The Future of Intelligent Agents in Cyberspace. *Journal of Interactive Marketing, 13*(3), 66-76.

Maiden, N. A. M., & Rugg, G. (1996). ACRE: selecting methods for requirements acquisition. *Software Engineering Journal, 11*(3), 183-192.

Mantel, S. J., & Rosegger, G. (1987). The role of third-parties in the diffusion of innovations: a survey. In R. Rothwell & J. Bessant (Eds.), *Innovation: Adaptation and Growth* (pp. 123-134). Amsterdam: Elsevier.

Maxwell, J. A. (1992). Understanding and validity in qualitative research. *Harvard educational review, 62*(3), 279-301.

Mayhew, D. J. (1991). *Principles and Guidelines in Software User Interface Design*. Upper Saddle River, NJ: Prentice Hall.

Mayring, P. (2008). *Qualitative Inhaltsanalyse: Grundlagen und Techniken*. Weinheim: Beltz.

McEvily, B., & Zaheer, A. (1999). Bridging ties: a source of firm heterogeneity in competitive capabilities. *Strategic Management Journal, 20*(12), 1133-1156.

McWilliam, G. (2000). Building stronger brands through online communities. *MIT Sloan Management Review, 41*(3), 43-54.

Van der Meulen, B., & Rip, A. (1998). Mediation in the Dutch science system. *Research Policy, 27*(8), 757-769.

Mich, L., Franch, M., & Gaio, L. (2003). Evaluating and designing Web site quality. *IEEE Multimedia, 10*(1), 34-43.

Miles, M. B., & Huberman, A. M. (1994). *Qualitative Data Analysis: An Expanded Sourcebook* (2nd ed.). Thousand Oaks, CA: Sage Publications.

Moeslein, K. M., Haller, J. B. A., & Bullinger, A. C. (2010). Open Evaluation: Ein IT-basierter Ansatz für die Bewertung innovativer Konzepte. *HMD - Praxis der Wirtschaftsinformatik, 273*, 21–34.

Moeslein, K. M., & Neyer, A.-K. (2009). Open Innovation: Grundlagen, Grenzen, Spannungsfelder. In A. Zerfass & K. M. Moeslein (Eds.), *Kommunikation als Erfolgsfaktor im Innovationsmanagement - Strategien im Zeitalter der Open Innovation* (pp. 85–103). Wiesbaden: Gabler.

Moon, J. Y., & Sproull, L. S. (2008). The Role of Feedback in Managing the Internet-Based Volunteer Work Force. *Information Systems Research, 19*(4), 494–515.

Morrison, P. D., Roberts, J. H., & Midgley, D. F. (2004). The nature of lead users and measurement of leading edge status. *Research Policy, 33*(2), 351–362.

Muniz Jr., A. M., & O'Guinn, T. C. (2001). Brand Communities. *Journal of Consumer Research, 27*, 412–432.

Nambisan, S. (2002). Designing virtual customer environments for new product development: Toward a theory. *Academy of Management Review, 27*(3), 392–413.

Neill, C. J., & Laplante, P. A. (2003). Requirements engineering: The state of the practice. *IEEE Software, 20*(6), 40–45.

Netcraft. (2012). June 2012 Web Server Survey | Netcraft. Retrieved June 26, 2012, from http://news.netcraft.com/archives/2012/06/06/june-2012-web-server-survey.html.

Neyer, A.-K., Bullinger, A. C., & Moeslein, K. M. (2009). Integrating inside and outside innovators: a sociotechnical systems perspective. *R&D Management, 39*(4), 410–419.

Nosek, J. T., & Palvia, P. (1990). Software maintenance management: Changes in the last decade. *Journal of Software Maintenance: Research and Practice, 2*(3), 157–174.

OECD (1991). The nature of innovation and the evolution of the productive system. *Technology and productivity - the challenge for economic policy* (pp. 303–314). Paris: OECD.

Oecking, C., & Degenhardt, A. (2011). Application Management 2.0. In F. Keuper, C. Oecking, & A. Degenhardt (Eds.), *Application Management* (pp. 4–29). Wiesbaden: Gabler.

Ogawa, S., & Piller, F. T. (2006). Reducing the Risks of New Product Development. *MIT Sloan Management Review, 47*(2), 65–71.

OGC (2002). *Application management* (4th ed.). Norwich: The Stationery Office.

OGC (2011). *ITIL Service operation* (2nd ed.). London: The Stationery Office.

Orlikowski, W. J., Yates, J., Okamura, K., & Fujimoto, M. (1995). Shaping Electronic Communication: The Metastructuring of Technology in the Context of Use. *Organization Science, 6*(4), 423–444.

O'Reilly, T. (2005). What is Web 2.0: Design patterns and business models for the next generation. *Communications & Strategies, 65*(1), 17–37.

Padgett, D. K. (1998). *Qualitative Methods in Social Work Research: Challenges and Rewards.* Thousand Oaks, CA: Sage Publications, Inc.

Pasmore, W., Francis, C., Haldeman, J., & Shani, A. (1982). Sociotechnical Systems: A North American Reflection on Empirical Studies of the Seventies. *Human Relations, 35*(12), 1179–1204.

Peters, T. J., & Waterman, R. H. (1982). *In Search of Excellence: Lessons from America's Best-Run Companies.* New York, NY: Harper and Row.

Pham, H. (1999, August 21). Software Reliability. *Wiley Encyclopedia of Electrical and Electronics Engineering.* Hoboken, NJ: Wiley Online Library.

Piller, F. T., & Walcher, D. (2006). Toolkits for idea competitions: a novel method to integrate users in new product development. *R&D Management, 36*(3), 307–318.

Piore, M. J. (2001). The emergent role of social intermediaries in the new economy. *Annals of Public and Cooperative Economics, 72*(3), 339–350.

Pittaway, L., Robertson, M., Munir, K., Denyer, D., & Neely, A. D. (2004). Networking and Innovation: A Systematic Review of the Evidence. *International Journal of Management Reviews, 5/6*(3&4), 137–168.

Pohl, K. (2008). *Requirements Engineering.* Heidelberg: Dpunkt.Verlag.

Powell, A., Piccoli, G., & Ives, B. (2004). Virtual teams: a review of current literature and directions for future research. *ACM SIGMIS Database, 35*(1), 6–36.

Prahalad, C. K., & Ramaswamy, V. (2004). *The Future of Competition: Co-Creating Unique Value With Customers.* Boston, MA: Harvard Business School Press.

Preece, J., & Maloney-Krichmar, D. (2006). Online Communities: Design, Theory, and Practice. *Journal of Computer-Mediated Communication, 10*(4).

Pressman, R. S. (1998). Can Internet-Based Applications Be Engineered. *IEEE Software, 15*(5), 104–110.

Provan, K. G., & Human, S. E. (1999). Organizational learning and the role of the network broker in small-firm manufacturing networks. In A. Grandori (Ed.), *Interfirm Networks: Organization and Industrial Competitiveness* (pp. 185–207). London: Routledge.

Quinn, J. B. (2000). Outsourcing Innovation: The New Engine of Growth. *Sloan Management Review, 41*(4), 13–28.

Reichwald, R., & Piller, F. (2009). *Interaktive Wertschöpfung - Open Innovation, Individualisierung und neue Formen der Arbeitsteilung* (2nd ed.). Wiesbaden: Gabler.

Rheingold, H. (2000). *The virtual community: Homesteading on the electronic frontier.* Cambridge, MA: The MIT Press.

Riedl, C., Blohm, I., Leimeister, J. M., & Krcmar, H. (2010). Rating scales for collective intelligence in innovation communities: Why quick and easy decision making does not get it right. *International Conference on Information Systems* (pp. 1–21). St. Louis, MO.

Robson, C. (2002). *Real world research.* Oxford: Blackwell Publishing.

Rocco, T. S., & Hatcher, T. (2011). *The Handbook of Scholarly Writing and Publishing.* San Francisco, CA: Jossey-Bass.

Rogers, E. M. (2003). *Diffusion of Innovations* (5th ed.). New York, NY: The Free Press.

Royce, W. (1970). Managing the development of large software systems. *Proceedings of IEEE WESCON* (pp. 328–338). Washington, DC.

Ruefli, T., Whinston, A., & Wiggins, R. R. (2001). The digital technological environment. In J. Wind & V. Mahajan (Eds.), *Digital Marketing: Global Strategies from the World's Leading Experts* (pp. 26–59). New York, NY: Wiley.

Rumbaugh, J., Jacobson, I., & Booch, G. (1999). *The Unified Software Development Process.* Boston, MA: Addison-Wesley.

Runeson, P., & Hoest, M. (2008). Guidelines for conducting and reporting case study research in software engineering. *Empirical Software Engineering, 14*(2), 131–164.

Sapsed, J., Grantham, A., & DeFillippi, R. (2007). A bridge over troubled waters: Bridging organisations and entrepreneurial opportunities in emerging sectors. *Research Policy, 36*(9), 1314 – 1334.

Sasankar, A. B., & Chavan, V. (2011). Survey of Software Life Cycle Models by Various Documented Standards. *International Journal of Computer Science & Technology, 2*(4), 137–144.

Sawhney, M., & Prandelli, E. (2000). Communities of Creation: Managing Distributed Innovation in Turbulent Markets. *California Management Review, 42*(4), 24-54.

Sawhney, M., Verona, G., & Prandelli, E. (2005). Collaborating to create: The Internet as a platform for customer engagement in product innovation. *Journal of Interactive Marketing, 19*(4), 4-17.

Scacchi, W. (2001). Process models in software engineering. In J. J. Marciniak (Ed.), *Encyclopedia of software engineering* (pp. 1-24). New York, NY: John Wiley & Sons.

Schnell, R., Hill, P. B., & Esser, E. (2008). *Methoden der empirischen Sozialforschung* (8th ed.). München: Oldenbourg Wissenschaftsverlag.

Schumpeter, J. A. (1934). *The theory of economic development: an inquiry into profits, capital, credit, interest, and the business cycle*. Cambridge, MA: Harvard University Press.

Schwaber, K. (2004). *Agile project management with Scrum*. Redmond: Microsoft Press.

Seaton, R. A. F., & Cordey-Hayes, M. (1993). The development and application of interactive models of industrial technology transfer. *Technovation, 13*(1), 45-53.

Shapiro, C., & Varian, H. R. (1999). *Information Rules: A Strategic Guide to the Network Economy*. Boston, MA: Harvard Business School Press.

Shaw, K. A. (2007). *Application lifecycle management for the enterprise*. San Mateo, CA: Serena.

Sherry, J. F., & Kozinets, R. V. (2001). Qualitative inquiry in marketing and consumer research. In D. Iacobucci (Ed.), *Kellogg on marketing* (pp. 165-194). New York, NY: John Wiley.

Shohet, S., & Prevezer, M. (1996). UK biotechnology: institutional linkages, technology transfer and the role of intermediaries. *R&D Management, 26*(3), 283-298.

Sieg, J. H., Wallin, M. W., & Von Krogh, G. (2010). Managerial challenges in open innovation: a study of innovation intermediation in the chemical industry. *R&D Management, 40*(3), 281-291.

Simon, H. (1957). *Administrative Behavior*. New York, NY: The Free Press.

Smith, A., Banzaert, A., & Susnowitz, S. (2003). The MIT ideas competition: promoting innovation for public service. *Proceedings of the 33rd IEEE Annual Conference on Frontiers in Education* (pp. 1-6). Boulder, CO.

Smith, C. (2002). The wholesale and retail markets of London, 1660–1840. *The Economic History Review, 55*(1), 31–50.

Smits, R. (2002). Innovation studies in the 21st century. *Technological Forecasting and Social Change, 69*(9), 861–883.

Soll, J. H. (2006). *Ideengenerierung mit Konsumenten im Internet.* Wiesbaden: Deutscher Universitätsverlag.

Sommerville, I. (2011). *Software engineering.* Boston, MA: Pearson.

Sousa, M. J. (1998). A Survey on the Software Maintenance Process. *Proceedings of the International Conference on Software Maintenance* (p. 265). Washington, D.C.

Spiegel Online. (2011). Soziale Netzwerke: Pril-Wettbewerb endet im PR-Debakel. Retrieved October 26, 2011, from http://www.spiegel.de/netzwelt/netzpolitik/soziale-netzwerke-pril-wettbewerb-endet-im-pr-debakel-a-763808.html.

Sproull, L., & Kiesler, S. (1991). *Connections. New ways of working in the networked organization.* Cambridge, MA: MIT Press.

Stake, R. E. (1995). *The art of case study research.* Thousand Oaks, CA: Sage Publications.

Stankiewicz, R. (1995). The role of the science and technology infrastructure in the development and diffusion of industrial automation in Sweden. In B. Carlsson (Ed.), *Technological Systems and Economic Performance: The Case of Factory Automation* (pp. 165–210). Kluwer: Dordrecht.

Stewart, J., & Hyysalo, S. (2008). Intermediaries, users and social learning in technological innovation. *International Journal of Innovation Management, 12*(3), 57–87.

Stone, D., Jarrett, C., Woodroffe, M., & Minocha, S. (2005). *User Interface Design and Evaluation.* San Francisco, CA: Morgan Kaufmann.

Strauss, A. L., & Corbin, J. M. (1998). *Basics of qualitative research: techniques and procedures for developing grounded theory* (2nd ed.). Thousand Oaks, CA: Sage Publications.

Surowiecki, J. (2005). *Wisdom of Crowds.* New York, NY: Abacus.

Takeuchi, H., & Nonaka, I. (1986). The new new product development game. *Harvard Business Review, 1*, 137–146.

Taylor, J. (2003). *Managing Information Technology Projects: Applying Project Management Strategies to Software, Hardware, and Integration Initiatives*. New York, NY: AMACOM.

Taylor, M., McWilliam, J., Forsyth, H., & Wade, S. (2002). Methodologies and website development: a survey of practice. *Information and Software Technology, 44*(6), 381–391.

Terwiesch, C., & Xu, Y. (2008). Innovation Contests, Open Innovation, and Multiagent Problem Solving. *Management Science, 54*(9), 1529–1543.

Tether, B., & Tajar, A. (2008). Beyond industry–university links: Sourcing knowledge for innovation from consultants, private research organisations and the public science-base. *Research Policy, 37*(6-7), 1079–1095.

Thayer, R. H. (2000). *Software Engineering Project Management* (2nd ed.). Danvers, MA: Wiley-IEEE Computer Society.

Tidd, J., & Bessant, J. (2009). *Managing Innovation: Integrating Technological, Market and Organizational Change*. West Sussex: John Wiley & Sons.

Tidwell, J. (2005). *Designing Interfaces*. Sebastopol, CA: O'Reilly.

Toubia, O. (2006). Idea generation, creativity, and incentives. *Marketing Science, 25*(5), 411–425.

Trinder, L., & Reynolds, S. (2000). *Evidence-based practice: a critical appraisal*. Oxford: Blackwell Science.

Trist, E. L. (1981). The evolution of socio-technical systems. *Occasional paper No. 2* (pp. 1–67). Toronto, ON: Ontario Ministry of Labour.

Trist, E. L., & Bamforth, K. W. (1951). Some Social and Psychological Consequences of the Longxwall Method of Coal-Getting: An Examination of the Psychological Situation and Defences of a Work Group in Relation to the Social Structure and Technological Content of the Work System. *Human Relations, 4*(1), 3–38.

Trist, E. L., Higgin, G. W., Murray, H., & Pollock, A. B. (1963). *Organizational choice: capabilities of groups at the coal face under changing technologies: the loss, re-discovery & transformation of a work tradition*. London: Travistock Publications.

Turpin, T., Garrett-Jones, S., & Rankin, N. (1996). Bricoleurs and boundary riders: managing basic research and innovation knowledge networks. *R&D Management, 26*(3), 267–282.

Vahs, D., & Burmester, R. (2005). *Innovationsmanagement. Von der Produktidee zur erfolgreichen Vermarktung* (3rd ed.). Stuttgart: Schäffer-Poeschel.

Ven, A. H. Van de. (2007). *Engaged Scholarship: A Guide for Organizational and Social Research*. New York, NY: Oxford University Press.

Verona, G., & Prandelli, E. (2002). A Dynamic Model of Customer Loyalty to Sustain Competitive Advantage on the Web. *European Management Journal, 20*(3), 299–309.

Verona, G., Prandelli, E., & Sawhney, M. (2006). Innovation and Virtual Environments: Towards Virtual Knowledge Brokers. *Organization Studies, 27*(6), 765–788.

Verworn, B., & Herstatt, C. (2000). Modelle des Innovationsprozesses: eine Einführung. *Arbeitspapier Nr. 6* (pp. 1–13). Hamburg: Technische Universität Hamburg-Harburg.

Walcher, D. (2007). *Der Ideenwettbewerb als Methode der aktiven Kundenintegration*. Wiesbaden: Gabler.

Watkins, D., & Horley, G. (1986). Transferring technology from large to small firms: the role of intermediaries. In T. Webb, T. Quince, & D. Watkins (Eds.), *Small Business Research* (pp. 215–251). Aldershot: Gower.

Wayland, R. E., & Cole, P. M. (1997). *Customer Connections: New Strategies for Growth*. Boston, MA: Harvard Business School Press.

Weischedel, B., & Huizingh, E. K. R. E. (2006). Website optimization with web metrics. *Proceedings of the 8th international conference on electronic commerce* (pp. 463–470). New York, NY.

Werbach, K. (2000). Syndication: the emerging model for business in the Internet era. *Harvard Business Review, 78*(3), 85–93.

Whitworth, B., Fjermestad, J., & Mahinda, E. (2006). The Web of System Performance. *Communications of the ACM, 49*(5), 93–100.

Wieringa, R., Maiden, N., Mead, N., & Rolland, C. (2005). Requirements engineering paper classification and evaluation criteria: a proposal and a discussion. *Requirements Engineering, 11*(1), 102–107.

Williams, D. (2011). *The Forrester Wave: Co-Creation Contest Vendors, Q3 2011*. Cambridge, MA: Forrester Research.

Williams, D., Gownder, J. P., Wiramihardja, L., & Corbett, A. E. (2010). *US Consumers Are Willing Co-Creators - Activate Engaged Consumers With Social Technologies To Build Better Products*. Cambridge, MA: Forrester Research.

Williams, R. L. (2000). Four Smart Ways To Run Online Communities. *Sloan Management Review, 41*(4), 81–91.

Winch, G. M., & Courtney, R. (2007). The Organization of Innovation Brokers: An International Review. *Technology Analysis & Strategic Management, 19*(6), 747–763.

Wood, P. (2002). Knowledge-intensive Services and Urban Innovativeness. *Urban Studies, 39*(5-6), 993–1002.

Yadav, S., Bravoco, R., Chatfield, A., & Rajkumar, T. M. (1988). Comparison of analysis techniques for information requirement determination. *Communications of the ACM, 31*(9), 1090–1097.

Yin, R. K. (2009). *Case study research: Design and methods*. Thousand Oaks, CA: Sage Publications.

Youtie, J., & Shapira, P. (2008). Building an innovation hub: A case study of the transformation of university roles in regional technological and economic development. *Research Policy, 37*(8), 1188–1204.

Zerdick, A., Schrape, K., Artope, A., Goldhammer, K., Heger, D. K., Lange, U. T., Vierkant, E., et al. (2001). *Die Internet-Ökonomie: Strategien für die digitale Wirtschaft (European Communication Council Report)* (3rd ed.). Berlin: Springer.

Zhang, Y., & Hiltz, S. R. (2003). Factors that influence online relationship development in a knowledge sharing community. *Proceedings of the Ninth American Conference on Information Systems* (pp. 410–417). Tampa, FL.

Zowghi, D., & Coulin, C. (2005). Requirements Elicitation: A Survey of Techniques, Approaches, and Tools. In A. Aurum & C. Wohlin (Eds.), *Engineering and Managing Software Requirements* (pp. 19–46). Heidelberg: Springer.

Springer Gabler RESEARCH

„Markt- und Unternehmensentwicklung / Markets and Organisations"
Herausgeber: Prof. Dr. Dres. h.c. Arnold Picot,
Prof. Dr. Prof. h.c. Dr. h.c. Ralf Reichwald, Prof. Dr. Egon Franck,
Prof. Dr. Kathrin Möslein
zuletzt erschienen:

Bastian Bansemir
Organizational Innovation Communities
2013. XVIII, 180 S., 14 Abb., 68 Tab., Br. € 59,95
ISBN 978-3-658-01301-1

Stefan H. Hallerstede
Managing the Lifecycle of Open Innovation Platforms
2013. XXIV, 249 S., 56 Abb., 59 Tab., Br. € 59,99
ISBN 978-3-658-02507-6

Kay H. Hofmann
Co-Financing Hollywood Film Productions with Outside Investors
An Economic Analysis of Principal Agent Relationships in the U.S. Motion Picture
Industry
2013. XVIII, 159 S., 10 Abb., 30 Tab., Br. € 59,95
ISBN 978-3-658-00786-7

Ralph Pfaller
IT-Outsourcing-Entscheidungen
Analyse von Einfluss- und Erfolgsfaktoren für auslagernde Unternehmen
2013. XX, 203 S., 31 Abb., 25 Tab., Br. € 49,95
ISBN 978-3-658-00714-0

Jessica Scheler
Driving Innovation in Service Organisations
A Study in the German Airport Industry
2013. XIX, 196 S., 16 Abb., 19 Tab., Br. € 49,95
ISBN 978-3-8349-3406-2

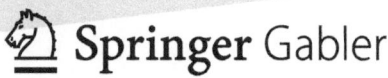

Springer Gabler

Änderungen vorbehalten. Stand: April 2013. Erhältlich im Buchhandel oder beim Verlag.
Abraham-Lincoln-Str. 46 . 65189 Wiesbaden . www.springer-gabler.de